Basic Math & Pre-Algebra Workbook

FOR DUMMIES®
A Wiley Brand

2nd Edition

by Mark Zegarelli

Basic Math & Pre-Algebra Workbook For Dummies®, 2nd Edition

Published by:
John Wiley & Sons, Inc.,
111 River Street,
Hoboken, NJ 07030-5774,
www.wiley.com

For general information on our other products and services, please contact our Customer Care Department within the U.S. at 877-762-2974, outside the U.S. at 317-572-3993, or fax 317-572-4002. For technical support, please visit www.wiley.com/techsupport.

Wiley publishes in a variety of print and electronic formats and by print-on-demand. Some material included with standard print versions of this book may not be included in e-books or in print-on-demand. If this book refers to media such as a CD or DVD that is not included in the version you purchased, you may download this material at http://booksupport.wiley.com. For more information about Wiley products, visit www.wiley.com.

Library of Congress Control Number: 2013954231

ISBN: 978-1-118-82804-5 (pbk); ISBN 978-1-118-82806-9 (ebk); ISBN 978-1-118-82830-4 (ebk)

Manufactured in the United States of America

10 9 8 7 6 5 4 3 2 1

Contents at a Glance

Table of Contents

Introduction

W hen you approach math right, it's almost always easier than you think. And a lot of the stuff that hung you up when you first saw it probably isn't all that scary after all. Lots of students feel they got lost somewhere along the way on the road between learning to count to ten and their first day in an algebra class — and this may be true whether you're 14 or 104. If this is you, don't worry. You're not alone, and help is right here!

Basic Math & Pre-Algebra Workbook For Dummies can give you the confidence and math skills you need to succeed in any math course you encounter on the way to algebra. One of the easiest ways to build confidence is to get experience working problems, allowing you to build those skills quickly. Everything in this book is designed to help clear the path on your math journey. Every section of every chapter contains a clear explanation of what you need to know, with plenty of practice problems and step-by-step solutions to every problem. Just grab a pencil, open this book to any page, and begin strengthening your math muscles!

About This Book

This book is for anyone who wants to improve his or her math skills. You may already be enrolled in a math class or preparing to register for one or simply studying on your own. In any case, practice makes perfect, and in this book you get plenty of practice solving a wide variety of math problems.

Each chapter covers a different topic in math: negative numbers, fractions, decimals, geometry, graphing, basic algebra — it's all here. In every section within a chapter, you find problems that allow you to practice a different skill. Each section features the following:

- ↳ A brief introduction to that section's topic
- ↳ An explanation of how to solve the problems in that section
- ↳ Sample questions with answers that show you all the steps to solving the problem
- ↳ Practice problems with space to work out your answer

Go ahead and write in this book — that's what it's for! When you've completed a problem or group of problems, flip to the end of the chapter. You'll find the correct answer followed by a detailed, step-by-step explanation of how to get there.

Although you can certainly work all the exercises in this book from beginning to end, you don't have to. Feel free to jump directly to whatever chapter has the type of problems you want to practice. When you've worked through enough problems in a section to your satisfaction, feel free to jump to a different section. If you find the problems in a section too difficult, flip back to an earlier section or chapter to practice the skills you need — just follow the cross-references.

Foolish Assumptions

You probably realize that the best way to figure out math is by doing it. You only want enough explanation to get down to business so you can put your math skills to work right away. If so, you've come to the right place. If you're looking for a more in-depth discussion, including tips on how all these math concepts fit into word problems, you may want to pick up the companion book, *Basic Math & Pre-Algebra For Dummies*.

I'm willing to bet my last dollar on earth that you're ready for this book. I assume only that you have some familiarity with the basics of the number system and the Big Four operations (adding, subtracting, multiplying, and dividing). To make sure that you're ready, take a look at these four arithmetic problems and see whether you can answer them:

$3 + 4 =$ _____

$10 - 8 =$ _____

$5 \times 5 =$ _____

$20 \div 2 =$ _____

If you can do these problems, you're good to go!

Icons Used in This Book

Throughout this book, I highlight some of the most important information with a variety of icons. Here's what they all mean:

This icon points out some of the most important pieces of information. Pay special attention to these details — you need to know them!

Tips show you a quick and easy way to do a problem. Try these tricks as you're solving the problems in that section.

Warnings are math booby traps that unwary students often fall into. Reading these bits carefully can help you avoid unnecessary heartache.

This icon highlights the example problems that show you techniques before you dive into the exercises.

Beyond the Book

In addition to the material in the print or e-book you're reading right now, this product also comes with some access-anywhere goodies on the web. Be sure to check out the free Cheat Sheet at www.dummies.com/cheatsheet/basicmathprealgebrawb for a set of quick reference notes including the order of operations, mathematical inequalities, basic algebra conventions, and more.

In addition, www.dummies.com/extras/basicmathprealgebrawb also contains related material on everything from converting between fractions and repeating decimals to great mathematicians from the last 2,500 years.

If you need more a detailed look at any of the concepts in this book, *Basic Math & Pre-Algebra For Dummies* will help you understand with crystal-clear explanations and plenty of examples. And if you want even more practice problems than are available here, *1,001 Practice Problems in Basic Math & Pre-Algebra For Dummies* provides lots more. Check them out!

Where to Go from Here

You can turn to virtually any page in this book and begin improving your math skills. Chapters 3 through 6 cover topics that tend to hang up math students: negative numbers, order of operations, factors and multiples, and fractions. A lot of what follows later in the book builds on these important early topics, so check them out. When you feel comfortable doing these types of problems, you have a real advantage in any math class.

Of course, if you already have a good handle on these topics, you can go anywhere you want (though you may still want to skim these chapters for some tips and tricks). My only advice is that you do the problems *before* reading the answer key!

And by all means, while you're at it, pick up *Basic Math & Pre-Algebra For Dummies,* which contains more-detailed explanations and a few extra topics not covered in this workbook. Used in conjunction, these two books can provide a powerful one-two punch to take just about any math problem to the mat.

Part I

Getting Started with Basic Math and Pre-Algebra

In this part. . .

- ✔ Understand place value.
- ✔ Use the Big Four operations: addition, subtraction, multiplication, and division.
- ✔ Calculate with negative numbers.
- ✔ Simplify expressions using the order of operations (PEMDAS).
- ✔ Work with factors and multiples.

Chapter 1

We've Got Your Numbers

In This Chapter

▶ Understanding how place value turns digits into numbers

▶ Rounding numbers to the nearest ten, hundred, or thousand

▶ Calculating with the Big Four operations: Adding, subtracting, multiplying, and dividing

▶ Getting comfortable with long division

*I*n this chapter, I give you a review of basic math, and I do mean basic. I bet you know a lot of this stuff already. So, consider this a trip down memory lane, a mini-vacation from whatever math you may be working on right now. With a really strong foundation in these areas, you'll find the chapters that follow a lot easier.

First, I discuss how the number system you're familiar with — called the *Hindu-Arabic number system* (or decimal numbers) — uses digits and place value to express numbers. Next, I show you how to round numbers to the nearest ten, hundred, or thousand.

After that, I discuss the Big Four operations: adding, subtracting, multiplying, and dividing. You see how to use the number line to make sense of all four operations. Then I give you practice doing calculations with larger numbers. To finish up, I make sure you know how to do long division both with and without a remainder.

Some math books use a dot (·) to indicate multiplication. In this book, I use the more familiar times sign (×).

Getting in Place with Numbers and Digits

The number system used most commonly throughout the world is the Hindu-Arabic number system. This system contains ten *digits* (also called *numerals*), which are symbols like the letters *A* through *Z*. I'm sure you're quite familiar with them:

$$1 \quad 2 \quad 3 \quad 4 \quad 5 \quad 6 \quad 7 \quad 8 \quad 9 \quad 0$$

Like letters of the alphabet, individual digits aren't very useful. When used in combination, however, these ten symbols can build numbers as large as you like using *place value*. Place value assigns each digit a greater or lesser value depending upon where it appears in a number. Each place in a number is ten times greater than the place to its immediate right.

Although the digit 0 adds no value to a number, it can act as a placeholder. When a 0 appears to the right of *at least one* nonzero digit, it's a placeholder. Placeholders are important for giving digits their proper place value. In contrast, when a 0 isn't to the right of any nonzero digit, it's a *leading zero*. Leading zeros are unnecessary and can be removed from a number.

Q. In the number 284, identify the ones digit, the tens digit, and the hundreds digit.

A. The ones digit is 4, the tens digit is 8, and the hundreds digit is 2.

Q. Place the number 5,672 in a table that shows the value of each digit. Then use this table and an addition problem to show how this number breaks down digit by digit.

A.

Millions	Hundred Thousands	Ten Thousands	Thousands	Hundreds	Tens	Ones
			5	6	7	2

The numeral 5 is in the thousands place, 6 is in the hundreds place, 7 is in the tens place, and 2 is in the ones place, so here's how the number breaks down:

5,000 + 600 + 70 + 2 = 5,672

Q. Place the number 040,120 in a table that shows the value of each digit. Then use this table to show how this number breaks down digit by digit. Which 0s are placeholders, and which are leading zeros?

A.

Millions	Hundred Thousands	Ten Thousands	Thousands	Hundreds	Tens	Ones
	0	4	0	1	2	0

The first 0 is in the hundred-thousands place, 4 is in the ten-thousands place, the next 0 is in the thousands place, 1 is in the hundreds place, 2 is in the tens place, and the last 0 is in the ones place, so

0 + 40,000 + 0 + 100 + 20 + 0 = 40,120

The first 0 is a leading zero, and the remaining 0s are placeholders.

1. In the number 7,359, identify the following digits:

   a. The ones digit

   b. The tens digit

   c. The hundreds digit

   d. The thousands digit

Solve It

---

2. Place the number 2,136 in a table that shows the value of each digit. Then use this table to show how this number breaks down digit by digit.

Millions	Hundred Thousands	Ten Thousands	Thousands	Hundreds	Tens	Ones

Solve It

---

3. Place the number 03,809 in a table that shows the value of each digit. Then use this table to show how this number breaks down digit by digit. Which 0 is a placeholder and which is a leading zero?

Millions	Hundred Thousands	Ten Thousands	Thousands	Hundreds	Tens	Ones

Solve It

---

4. Place the number 0,450,900 in a table that shows the value of each digit. Then use this table to show how this number breaks down digit by digit. Which 0s are placeholders and which are leading zeros?

Millions	Hundred Thousands	Ten Thousands	Thousands	Hundreds	Tens	Ones

Solve It

Rollover: Rounding Numbers Up and Down

Rounding numbers makes long numbers easier to work with. To round a two-digit number to the nearest ten, simply increase it or decrease it to the nearest number that ends in 0:

- When a number ends in 1, 2, 3, or 4, bring it down; in other words, keep the tens digit the same and turn the ones digit into a 0.

- When a number ends in 5, 6, 7, 8, or 9, bring it up; add 1 to the tens digit and turn the ones digit into a 0.

To round a number with more than two digits to the nearest ten, use the same method, focusing only on the ones and tens digits.

After you understand how to round a number to the nearest ten, rounding a number to the nearest hundred, thousand, or beyond is easy. Focus only on two digits: The digit in the place you're rounding to and the digit to its immediate right, which tells you whether to round up or down. All the digits to the right of the number you're rounding to change to 0s.

Occasionally when you're rounding a number up, a small change to the ones and tens digits affects the other digits. This is a lot like when the odometer in your car rolls a bunch of 9s over to 0s, such as when you go from 11,999 miles to 12,000 miles.

Q. Round the numbers 31, 58, and 95 to the nearest ten.

A. 30, 60, and 100.

The number 31 ends in 1, so round it down:

$31 \rightarrow 30$

The number 58 ends in 8, so round it up:

$58 \rightarrow 60$

The number 95 ends in 5, so round it up:

$95 \rightarrow 100$

Q. Round the numbers 742, 3,820, and 61,225 to the nearest ten.

A. 740, 3,820, and 61,230.

The number 742 ends in 2, so round it down:

$7\underline{42} \rightarrow 7\underline{40}$

The number 3,820 already ends in 0, so no rounding is needed:

$3,8\underline{20} \rightarrow 3,8\underline{20}$

The number 61,225 ends in 5, so round it up:

$61,2\underline{25} \rightarrow 61,2\underline{30}$

5. Round these two-digit numbers to the nearest ten:

   a. 29

   b. 43

   c. 75

   d. 97

Solve It

6. Round these numbers to the nearest ten:

   a. 164

   b. 765

   c. 1,989

   d. 9,999,995

Solve It

7. Round these numbers to the nearest hundred:

   a. 439

   b. 562

   c. 2,950

   d. 109,974

Solve It

8. Round these numbers to the nearest thousand:

   a. 5,280

   b. 77,777

   c. 1,234,567

   d. 1,899,999

Solve It

Using the Number Line with the Big Four

The *number line* is just a line with numbers marked off at regular intervals. You probably saw your first number line when you were learning how to count to ten. In this section, I show you how to use this trusty tool to perform the Big Four operations (adding, subtracting, multiplying, and dividing) on relatively small numbers.

The number line can be a useful tool for adding and subtracting small numbers:

- ✔ When you add, move *up* the number line, to the right.
- ✔ When you subtract, move *down* the number line, to the left.

To multiply on the number line, start at 0 and count by the *first number* in the problem as many times as indicated by the *second number*.

To divide on the number line, first block off a segment of the number line from 0 to the *first number* in the problem. Then divide this segment evenly into the number of pieces indicated by the *second number*. The length of each piece is the answer to the division.

Q. Add 6 + 7 on the number line.

A. **13.** The expression 6 + 7 means *start at 6, up 7,* which brings you to 13 (see Figure 1-1).

Figure 1-1
Adding
6 + 7 = 13 on
the number
line.

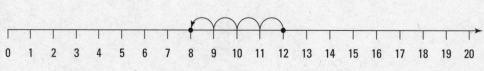

Q. Subtract 12 – 4 on the number line.

A. **8.** The expression 12 – 4 means *start at 12, down 4,* which brings you to 8 (see Figure 1-2).

Figure 1-2
Subtracting
12 – 4 = 8 on
the number
line.

Q. Multiply 2 × 5 on the number line.

A. **10.** Starting at 0, count by twos a total of five times, which brings you to 10 (see Figure 1-3).

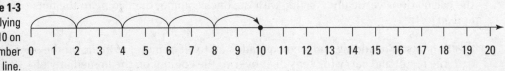

Figure 1-3
Multiplying
2 × 5 = 10 on
the number
line.

Q. Divide 12 ÷ 3 on the number line.

A. **4.** Block off the segment of the number line from 0 to 12. Now divide this segment evenly into three smaller pieces, as shown in Figure 1-4. Each of these pieces has a length of 4, so this is the answer to the problem.

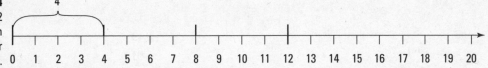

Figure 1-4
Dividing 12
÷ 3 = 4 on
the number
line.

---

9. Add the following numbers on the number line:

a. $4 + 7 = ?$

b. $9 + 8 = ?$

c. $12 + 0 = ?$

d. $4 + 6 + 1 + 5 = ?$

Solve It

10. Subtract the following numbers on the number line:

a. $10 - 6 = ?$

b. $14 - 9 = ?$

c. $18 - 18 = ?$

d. $9 - 3 + 7 - 2 + 1 = ?$

Solve It

---

11. Multiply the following numbers on the number line:

a. 2×7

b. 7×2

c. 4×3

d. 6×1

e. 6×0

f. 0×10

Solve It

12. Divide the following numbers on the number line:

a. $8 \div 2 = ?$

b. $15 \div 5 = ?$

c. $18 \div 3 = ?$

d. $10 \div 10 = ?$

e. $7 \div 1 = ?$

f. $0 \div 2 = ?$

Solve It

The Column Lineup: Adding and Subtracting

To add or subtract large numbers, stack the numbers on top of each other so that all similar digits (ones, tens, hundreds, and so forth) form columns. Then work from right to left. Do the calculations vertically, starting with the ones column, then going to the tens column, and so forth:

- ✔ When you're adding and a column adds up to 10 or more, write down the ones digit of the result and carry the tens digit over to the column on the immediate left.

- ✔ When you're subtracting and the top digit in a column is less than the bottom digit, borrow from the column on the immediate left.

Q. Add 35 + 26 + 142.

A. **203.** Stack the numbers and add the columns from right to left:

$$
\begin{array}{r}
^{11} \\
35 \\
26 \\
+142 \\
\hline
203
\end{array}
$$

Notice that when I add the ones column (5 + 6 + 2 = 13), I write the 3 below this column and carry the 1 over to the tens column. Then, when I add the tens column (1 + 3 + 2 + 4 = 10), I write the 0 below this column and carry the 1 over to the hundreds column.

Q. Subtract 843 – 91.

A. **752.** Stack the numbers and subtract the columns from right to left:

$$
\begin{array}{r}
^{1} \\
^{7}\cancel{8}43 \\
-91 \\
\hline
752
\end{array}
$$

When I try to subtract the tens column, 4 is less than 9, so I borrow 1 from the hundreds column, changing the 8 to 7. Then I place this 1 above the 4, changing it to 14. Now I can subtract 14 – 9 = 5.

13. Add 129 + 88 + 35.

Solve It

14. Find the following sum: 1,734 + 620 + 803 + 32 = ?

Solve It

15. Subtract 419 – 57.

Solve It

16. Subtract 41,024 – 1,786.

Solve It

Multiplying Multiple Digits

To multiply large numbers, stack the first number on top of the second. Then multiply each digit of the bottom number, from right to left, by the top number. In other words, first multiply the top number by the ones digit of the bottom number. Then write down a 0 as a placeholder and multiply the top number by the tens digit of the bottom number. Continue the process, adding placeholders and multiplying the top number by the next digit in the bottom number.

When the result is a two-digit number, write down the ones digit and carry the tens digit to the next column. After multiplying the next two digits, add the number you carried over.

Add the results to obtain the final answer.

Q. Multiply 742 × 136.

A. **100,912.** Stack the first number on top of the second:

$$
\begin{array}{r}
742 \\
\times 136 \\
\end{array}
$$

Now multiply 6 by every number in 742, starting from the right. Because 2 × 6 = 12, a two-digit number, you write down the 2 and carry the 1 to the tens column. In the next column, you multiply 4 × 6 = 24, and add the 1 you carried over, giving you a total of 25. Write down the 5, and carry the 2 to the hundreds column. Multiply 7 × 6 = 42, and add the 2 you carried over, giving you 44:

$$
\begin{array}{r}
{\scriptstyle 2\,1} \\
742 \\
\times 136 \\
\hline
4452 \\
\end{array}
$$

Next, write down a 0 all the way to the right in the row below the one that you just wrote. Multiply 3 by every number in 742, starting from the right and carrying when necessary:

$$
\begin{array}{r}
{\scriptstyle 1} \\
742 \\
\times 136 \\
\hline
4452 \\
22260 \\
\end{array}
$$

Write down two 0s all the way to the right of the row below the one that you just wrote. Repeat the process with 1:

$$
\begin{array}{r}
742 \\
\times 136 \\
\hline
4452 \\
22260 \\
74200 \\
\end{array}
$$

To finish, add up the results:

$$
\begin{array}{r}
742 \\
\times 136 \\
\hline
4452 \\
22260 \\
74200 \\
\hline
100912
\end{array}
$$

So $742 \times 136 = 100{,}912$.

---

17. Multiply 75×42.

Solve It

18. What's 136×84?

Solve It

---

19. Solve $1{,}728 \times 405$.

Solve It

20. Multiply $8{,}912 \times 767$.

Solve It

Cycling through Long Division

To divide larger numbers, use *long division*. Unlike the other Big Four operations, long division moves from left to right. For each digit in the *divisor* (the number you're dividing), you complete a cycle of division, multiplication, and subtraction.

In some problems, the number at the very bottom of the problem isn't a 0. In these cases, the answer has a *remainder,* which is a leftover piece that needs to be accounted for. In those cases, you write *r* followed by whatever number is left over.

Q. Divide 956 ÷ 4.

A. **239.** Start off by writing the problem like this:

$$4\overline{)956}$$

To begin, ask how many times 4 goes into 9 — that is, what's 9 ÷ 4? The answer is 2 (with a little left over), so write 2 directly above the 9. Now multiply 2 × 4 to get 8, place the answer directly below the 9, and draw a line beneath it:

$$\begin{array}{r} 2 \\ 4\overline{)956} \\ 8 \end{array}$$

Subtract 9 – 8 to get 1. (**Note:** After you subtract, the result should be less than the divisor (in this problem, the divisor is 4). Then bring down the next number (5) to make the new number 15.

$$\begin{array}{r} 2 \\ 4\overline{)956} \\ -8 \\ \hline 15 \end{array}$$

These steps are one complete cycle. To complete the problem, you just need to repeat them. Now ask how many times 4 goes into 15 — that is, what's 15 ÷ 4? The answer is 3 (with a little left over). So write the 3 above the 5, and then multiply 3 × 4 to get 12. Write the answer under 15.

$$\begin{array}{r} 23 \\ 4\overline{)956} \\ -8 \\ \hline 15 \\ -12 \end{array}$$

Subtract 15 – 12 to get 3. Then bring down the next number (6) to make the new number 36.

$$\begin{array}{r} 23 \\ 4\overline{)956} \\ -8 \\ \hline 15 \\ -12 \\ \hline 36 \end{array}$$

Another cycle is complete, so begin the next cycle by asking how many times 4 goes into 36 — that is, what's 36 ÷ 4? The answer this time is 9. Write down 9 above the 6, multiply 9 × 4, and place this below the 36.

$$\begin{array}{r} 239 \\ 4\overline{)956} \\ -8 \\ \hline 15 \\ -12 \\ \hline 36 \\ 36 \end{array}$$

Now subtract 36 – 36 = 0. Because you have no more numbers to bring down, you're finished, and the answer (that is, the *quotient*) is the very top number of the problem:

$$\begin{array}{r} 239 \\ 4\overline{)956} \\ -8 \\ \hline 15 \\ -12 \\ \hline 36 \\ -36 \\ \hline 0 \end{array}$$

Q. Divide 3,042 ÷ 5.

A. **608 r 2.** Start off by writing the problem like this:

5)‾3042‾

To begin, ask how many times 5 goes into 3. The answer is 0 — because 5 doesn't go into 3 — so write a 0 above the 3. Now you need to ask the same question using the first *two* digits of the divisor: How many times does 5 go into 30 — that is, what's 30 ÷ 5? The answer is 6, so place the 6 over the 0. Here's how to complete the first cycle:

```
    06
5)3042
  −30
────
   04
```

Next, ask how many times 5 goes into 4. The answer is 0 — because 5 doesn't go into 4 — so write a 0 above the 4. Now bring down the next number (2), to make the number 42:

```
   060
5)3042
  −30
────
   042
```

Ask how many times 5 goes into 42 — that is, what's 42 ÷ 5? The answer is 8 (with a little bit left over), so complete the cycle as follows:

```
  0608     ← quotient
5)3042
  −30
────
   042
  −40
────
     2     ← remainder
```

Because you have no more numbers to bring down, you're finished. The answer (quotient) is at the top of the problem (you can drop the leading 0), and the remainder is at the bottom of the problem. So 3,042 ÷ 5 = 608 with a remainder of 2. To save space, write this answer as 608 r 2.

21. Divide 741 ÷ 3.

Solve It

22. Evaluate 3,245 ÷ 5.

Solve It

23. Figure out 91,390 ÷ 8.

Solve It

24. Find 792,541 ÷ 9.

Solve It

Solutions to We've Got Your Numbers

The following are the answers to the practice questions presented in this chapter.

1 Identify the ones, tens, hundreds, and thousands digit in the number 7,359.

    a. 9 is the ones digit.

    b. 5 is the tens digit.

    c. 3 is the hundreds digit.

    d. 7 is the thousands digit.

2 2,000 + 100 + 30 + 6 = 2,136

Millions	Hundred Thousands	Ten Thousands	Thousands	Hundreds	Tens	Ones
			2	1	3	6

3 0 + 3,000 + 800 + 0 + 9 = 3,809. The first 0 is the leading zero, and the second 0 is the placeholder.

Millions	Hundred Thousands	Ten Thousands	Thousands	Hundreds	Tens	Ones
		0	3	8	0	9

4 0 + 400,000 + 50,000 + 0 + 900 + 0 + 0 = 0,450,900. The first 0 is a leading zero, and the remaining three 0s are placeholders.

Millions	Hundred Thousands	Ten Thousands	Thousands	Hundreds	Tens	Ones
0	4	5	0	9	0	0

5 Round to the nearest ten:

    a. 29 → **30.** The ones digit is 9, so round up.

    b. 43 → **40.** The ones digit is 3, so round down.

    c. 75 → **80.** The ones digit is 5, so round up.

    d. 97 → **100.** The ones digit is 7, so round up, rolling 9 over.

6 Round to the nearest ten:

    a. 16<u>4</u> → **160.** The ones digit is 4, so round down.

    b. 76<u>5</u> → **770.** The ones digit is 5, so round up.

    c. 1,98<u>9</u> → **1,990.** The ones digit is 9, so round up.

    d. 9,999,99<u>5</u> → **10,000,000.** The ones digit is 5, so round up, rolling all of the 9s over.

7 Focus on the hundreds and tens digits to round to the nearest hundred.

    a. 4<u>3</u>9 → **400.** The tens digit is 3, so round down.

    b. 5<u>6</u>2 → **600.** The tens digit is 6, so round up.

    c. 2,9<u>5</u>0 → **3,000.** The tens digit is 5, so round up.

    d. 109,<u>9</u>74 → **110,000.** The tens digit is 7, so round up, rolling over all the 9s.

8 Focus on the thousands and hundreds digits to round to the nearest thousand.

   a. 5,280 → **5,000.** The hundreds digit is 2, so round down.

   b. 77,777 → **78,000.** The hundreds digit is 7, so round up.

   c. 1,234,567 → **1,235,000.** The hundreds digit is 5, so round up.

   d. 1,899,999 → **1,900,000.** The hundreds digit is 9, so round up, rolling over all the 9s to the left.

9 Add on the number line.

   a. 4 + 7 = **11.** The expression 4 + 7 means *start at 4, up 7*, which brings you to 11.

   b. 9 + 8 = **17.** The expression 9 + 8 means *start at 9, up 8*, which brings you to 17.

   c. 12 + 0 = **12.** The expression 12 + 0 means *start at 12, up 0*, which brings you to 12.

   d. 4 + 6 + 1 + 5 = **16.** The expression 4 + 6 + 1 + 5 means *start at 4, up 6, up 1, up 5*, which brings you to 16.

10 Subtract on the number line.

   a. 10 − 6 = **4.** The expression 10 − 6 means *start at 10, down 6*, which brings you to 4.

   b. 14 − 9 = **5.** The expression 14 − 9 means *start at 14, down 9*, which brings you to 5.

   c. 18 − 18 = **0.** The expression 18 − 18 means *start at 18, down 18*, which brings you to 0.

   d. 9 − 3 + 7 − 2 + 1 = **12.** The expression 9 − 3 + 7 − 2 + 1 means *start at 9, down 3, up 7, down 2, up 1*, which brings you to 12.

11 Multiply on the number line.

   a. 2 × 7 = **14.** Starting at 0, count by twos a total of seven times, which brings you to 14.

   b. 7 × 2 = **14.** Starting at 0, count by sevens a total of two times, which brings you to 14.

   c. 4 × 3 = **12.** Starting at 0, count by fours a total of three times, which brings you to 12.

   d. 6 × 1 = **6.** Starting at 0, count by sixes one time, which brings you to 6.

   e. 6 × 0 = **0.** Starting at 0, count by sixes zero times, which brings you to 0.

   f. 0 × 10 = **0.** Starting at 0, count by zeros a total of ten times, which brings you to 0.

12 Divide on the number line.

   a. 8 ÷ 2 = **4.** Block off a segment of the number line from 0 to 8. Now divide this segment evenly into two smaller pieces. Each of these pieces has a length of 4, so this is the answer to the problem.

   b. 15 ÷ 5 = **3.** Block off a segment of the number line from 0 to 15. Divide this segment evenly into five smaller pieces. Each of these pieces has a length of 3, so this is the answer to the problem.

   c. 18 ÷ 3 = **6.** Block off a segment of the number line from 0 to 18 and divide this segment evenly into three smaller pieces. Each piece has a length of 6, the answer to the problem.

   d. 10 ÷ 10 = **1.** Block off a segment of the number line from 0 to 10 and divide this segment evenly into ten smaller pieces. Each of these pieces has a length of 1.

   e. 7 ÷ 1 = **7.** Block off a segment of the number line from 0 to 7 and divide this segment evenly into 1 piece (that is, don't divide it at all). This piece still has a length of 7.

   f. 0 ÷ 2 = **0.** Block off a segment of the number line from 0 to 0. The length of this segment is 0, so it can't get any smaller. This shows you that 0 divided by *any* number is 0.

13 **252**

$\overset{1\ 2}{}$
129
88
+35
———
252

14 **3,189**

$\overset{2}{}$
1734
620
803
+ 32
———
3189

15 **362**

$\overset{3\ 1}{\cancel{4}19}$
−57
———
362

16 **39,238**

$\overset{3\ 10\ 9\ 11\ 1}{\cancel{4}1\cancel{0}2\cancel{4}}$
− 1786
————
39238

17 **3,150**

75
×42
———
150
3000
———
3150

18 **11,424**

136
×84
———
544
10880
————
11424

19 **699,840**

1728
×405
———
8640
00000
691200
————
699840

20 **6,835,504**

8912
×767
———
62384
534720
6238400
————
6835504

21 **247**

$\overset{247}{3)\overline{741}}$
−6
——
14
−12
——
21
−21
——
0

22 **649**

$\overset{0649}{5)\overline{3245}}$
−30
——
24
−20
——
45
−45
——
0

23 **11,423 r 6**

$$
\begin{array}{r}
11423 \\
8\overline{)91390} \\
-8 \\
\hline
11 \\
-8 \\
\hline
33 \\
-32 \\
\hline
19 \\
-16 \\
\hline
30 \\
-24 \\
\hline
6
\end{array}
$$

24 **88,060 r 1**

$$
\begin{array}{r}
088060 \\
9\overline{)792541} \\
-72 \\
\hline
72 \\
-72 \\
\hline
054 \\
-54 \\
\hline
01 \\
-0 \\
\hline
1
\end{array}
$$

Chapter 2

Smooth Operators: Working with the Big Four Operations

The Big Four operations (adding, subtracting, multiplying, and dividing) are basic stuff, but they're really pretty versatile tools. In this chapter, I show you that the Big Four are really two pairs of inverse operations — that is, operations that undo each other. You also discover how the commutative property allows you to rearrange numbers in an expression. And most important, you find out how to rewrite equations in alternative forms that allow you to solve problems more easily.

Next, I show you how to use parentheses to group numbers and operations together. You discover how the associative property ensures that, in certain cases, the placement of parentheses doesn't change the answer to a problem. You also work with four types of inequalities: >, <, ≠, and ≈. Finally, I show you that raising a number to a power is a shortcut for multiplication and explain how to find the square root of a number.

Switching Things Up with Inverse Operations and the Commutative Property

The Big Four operations are actually two pairs of *inverse operations,* which means the operations can undo each other:

✔ **Addition and subtraction:** Subtraction undoes addition. For example, if you start with 3 and add 4, you get 7. Then, when you subtract 4, you undo the original addition and arrive back at 3:

$$3 + 4 = 7 \rightarrow 7 - 4 = 3$$

This idea of inverse operations makes a lot of sense when you look at the number line. On a number line, $3 + 4$ means *start at 3, up 4.* And $7 - 4$ means *start at 7, down 4.* So when you add 4 and then subtract 4, you end up back where you started.

✔ **Multiplication and division:** Division undoes multiplication. For example, if you start with 6 and multiply by 2, you get 12. Then, when you divide by 2, you undo the original multiplication and arrive back at 6:

$$6 \times 2 = 12 \rightarrow 12 \div 2 = 6$$

The *commutative property of addition* tells you that you can change the order of the numbers in an addition problem without changing the result, and the *commutative property of multiplication* says you can change the order of the numbers in a multiplication problem without changing the result. For example,

$$2 + 5 = 7 \rightarrow 5 + 2 = 7$$

$$3 \times 4 = 12 \rightarrow 4 \times 3 = 12$$

Through the commutative property and inverse operations, every equation has four alternative forms that contain the same information expressed in slightly different ways. For example, $2 + 3 = 5$ and $3 + 2 = 5$ are alternative forms of the same equation but tweaked using the commutative property. And $5 - 3 = 2$ is the inverse of $2 + 3 = 5$. Finally, $5 - 2 = 3$ is the inverse of $3 + 2 = 5$.

You can use alternative forms of equations to solve fill-in-the-blank problems. As long as you know two numbers in an equation, you can always find the remaining number. Just figure out a way to get the blank to the other side of the equal sign:

✔ When the *first* number is missing in any problem, use the inverse to turn the problem around:

$$\underline{\hspace{2cm}} + 6 = 10 \rightarrow 10 - 6 = \underline{\hspace{2cm}}$$

✔ When the *second* number is missing in an addition or multiplication problem, use the commutative property and then the inverse:

$$9 + \underline{\hspace{2cm}} = 17 \rightarrow \underline{\hspace{2cm}} + 9 = 17 \rightarrow 17 - 9 = \underline{\hspace{2cm}}$$

✔ When the *second* number is missing in a subtraction or multiplication problem, just switch around the two values that are next to the equal sign (that is, the blank and the equal sign):

$$15 - \underline{\hspace{2cm}} = 8 \rightarrow 15 - 8 = \underline{\hspace{2cm}}$$

Q. What's the inverse equation to $16 - 9 = 7$?

A. **7 + 9 = 16.** In the equation $16 - 9 = 7$, you start at 16 and subtract 9, which brings you to 7. The inverse equation undoes this process, so you start at 7 and add 9, which brings you back to 16:

$$16 - 9 = 7 \rightarrow 7 + 9 = 16$$

Q. What's the inverse equation to $6 \times 7 = 42$?

A. **42 ÷ 7 = 6.** In the equation $6 \times 7 = 42$, you start at 6 and multiply by 7, which brings you to 42. The inverse equation undoes this process, so you start at 42 and divide by 7, which brings you back to 6:

$$6 \times 7 = 42 \rightarrow 42 \div 7 = 6$$

Q. Use inverse operations and the commutative property to find three alternative forms of the equation $7 - 2 = 5$.

A. **$5 + 2 = 7$, $2 + 5 = 7$, and $7 - 5 = 2$.** First, use inverse operations to change subtraction to addition:

$$7 - 2 = 5 \rightarrow 5 + 2 = 7$$

Now use the commutative property to change the order of this addition:

$$5 + 2 = 7 \rightarrow 2 + 5 = 7$$

Finally, use inverse operations to change addition to subtraction:

$$2 + 5 = 7 \rightarrow 7 - 5 = 2$$

Q. Solve this problem by filling in the blank: $16 + \underline{\hspace{2cm}} = 47$.

A. **31.** First, use the commutative property to reverse the addition:

$$16 + \underline{\hspace{2cm}} = 47 \rightarrow$$
$$\underline{\hspace{2cm}} + 16 = 47$$

Now use inverse operations to change the problem from addition to subtraction:

$$\underline{\hspace{2cm}} + 16 = 47 \rightarrow 47 - 16 =$$
$$\underline{\hspace{2cm}}$$

At this point, you can solve the problem by subtracting $47 - 16 = 31$.

Q. Fill in the blank: $\underline{\hspace{2cm}} \div 3 = 13$.

A. **39.** Use inverse operations to turn the problem from division to multiplication:

$$\underline{\hspace{2cm}} \div 3 = 13 \rightarrow 13 \times 3 =$$
$$\underline{\hspace{2cm}}$$

Now you can solve the problem by multiplying $13 \times 3 = 39$.

Q. Fill in the blank: $64 - \underline{\hspace{2cm}} = 15$.

A. **49.** Switch around the last two numbers in the problem:

$$64 - \underline{\hspace{2cm}} = 15 \rightarrow 64 - 15 =$$
$$\underline{\hspace{2cm}}$$

Now you can solve the problem by subtracting $64 - 15 = 49$.

1. Using inverse operations, write down an alternative form of each equation:

    a. 8 + 9 = 17

    b. 23 − 13 = 10

    c. 15 × 5 = 75

    d. 132 ÷ 11 = 12

Solve It

2. Use the commutative property to write down an alternative form of each equation:

    a. 19 + 35 = 54

    b. 175 + 88 = 263

    c. 22 × 8 = 176

    d. 101 × 99 = 9,999

Solve It

3. Use inverse operations and the commutative property to find all three alternative forms for each equation:

    a. 7 + 3 = 10

    b. 12 − 4 = 8

    c. 6 × 5 = 30

    d. 18 ÷ 2 = 9

Solve It

4. Fill in the blanks in each question:

    a. _____ − 74 = 36

    b. _____ × 7 = 105

    c. 45 + _____ = 132

    d. 273 − _____ = 70

    e. 8 × _____ = 648

    f. 180 ÷ _____ = 9

Solve It

Getting with the In-Group: Parentheses and the Associative Property

Parentheses group operations together, telling you to do any operations inside a set of parentheses *before* you do operations outside of it. Parentheses can make a big difference in the result you get when solving a problem, especially in a problem with mixed operations. In two important cases, however, moving parentheses doesn't change the answer to a problem:

✔ The *associative property of addition* says that when every operation is addition, you can group numbers however you like and choose which pair of numbers to add first; you can move parentheses without changing the answer.

✔ The *associative property of multiplication* says you can choose which pair of numbers to multiply first, so when every operation is multiplication, you can move parentheses without changing the answer.

Taken together, the associative property and the commutative property (which I discuss in the preceding section) allow you to completely rearrange all the numbers in any problem that's either all addition or all multiplication.

Q. What's $(21 - 6) \div 3$? What's $21 - (6 \div 3)$?

A. **5 and 19.** To calculate $(21 - 6) \div 3$, first do the operation inside the parentheses — that is, $21 - 6 = 15$:

$$(21 - 6) \div 3 = 15 \div 3$$

Now finish the problem by dividing: $15 \div 3 = 5$.

To solve $21 - (6 \div 3)$, first do the operation inside the parentheses — that is, $6 \div 3 = 2$:

$$21 - (6 \div 3) = 21 - 2$$

Finish up by subtracting $21 - 2 = 19$. Notice that the placement of the parentheses changes the answer.

Q. Solve $1 + (9 + 2)$ and $(1 + 9) + 2$.

A. **12 and 12.** To solve $1 + (9 + 2)$, first do the operation inside the parentheses — that is, $9 + 2 = 11$:

$$1 + (9 + 2) = 1 + 11$$

Finish up by adding $1 + 11 = 12$.

To solve $(1 + 9) + 2$, first do the operation inside the parentheses — that is, $1 + 9 = 10$:

$$(1 + 9) + 2 = 10 + 2$$

Finish up by adding $10 + 2 = 12$. Notice that the only difference between the two problems is the placement of the parentheses, but because both operations are addition, moving the parentheses doesn't change the answer.

Q. Solve $2 \times (4 \times 3)$ and $(2 \times 4) \times 3$.

A. **24 and 24.** To solve $2 \times (4 \times 3)$, first do the operation inside the parentheses — that is, $4 \times 3 = 12$:

$$2 \times (4 \times 3) = 2 \times 12$$

Finish by multiplying $2 \times 12 = 24$.

To solve $(2 \times 4) \times 3$, first do the operation inside the parentheses — that is, $2 \times 4 = 8$:

$$(2 \times 4) \times 3 = 8 \times 3$$

Finish by multiplying $8 \times 3 = 24$. No matter how you group the multiplication, the answer is the same.

Q. Solve $41 \times 5 \times 2$.

A. **410.** The last two numbers are small, so place parentheses around these numbers:

$$41 \times 5 \times 2 = 41 \times (5 \times 2)$$

First, do the multiplication inside the parentheses:

$$41 \times (5 \times 2) = 41 \times 10$$

Now you can easily multiply $41 \times 10 = 410$.

5. Find the value of $(8 \times 6) + 10$.

Solve It

6. Find the value of $123 \div (145 - 144)$.

Solve It

7. Solve the following two problems:

   a. $(40 \div 2) + 6 = ?$

   b. $40 \div (2 + 6) = ?$

Do the parentheses make a difference in the answers?

Solve It

8. Solve the following two problems:

   a. $(16 + 24) + 19$

   b. $16 + (24 + 19)$

Do the parentheses make a difference in the answers?

Solve It

9. Solve the following two problems:

a. $(18 \times 25) \times 4$

b. $18 \times (25 \times 4)$

Do the parentheses make a difference in the answers?

Solve It

10. Find the value of $93,769 \times 2 \times 5$. (***Hint:*** Use the associative property for multiplication to make the problem easier.)

Solve It

Becoming Unbalanced: Inequalities

When numbers aren't equal in value, you can't use the equal sign (=) to turn them into an equation. Instead, you use a variety of other symbols to turn them into an *inequality:*

✔ The symbol > means *is greater than,* and the symbol < means *is less than:*

  6 > 3 means 6 *is greater than* 3.

  7 < 10 means 7 *is less than* 10.

If you're not sure whether to use > or <, remember that the big open mouth of the symbol always faces the larger number. For example, 5 < 7, but 7 > 5.

✔ The symbol ≠ means *doesn't equal.* It's not as useful as > or < because it doesn't tell you whether a number is greater than or less than another number. Mostly, ≠ points out a mistake or inaccuracy in a pre-algebra calculation.

✔ The symbol ≈ means *approximately equals.* You use it when you're rounding numbers and estimating solutions to problems — that is, when you're looking for an answer that's close enough but not exact. The symbol ≈ allows you to make small adjustments to numbers to make your work easier. (See Chapter 1 for more on estimating and rounding.)

Q. Place the correct symbol (=, >, or <) in the blank: 2 + 2 _____ 5.

A. **<.** Because 2 + 2 = 4 and 4 is less than 5, use the symbol that means *is less than*.

Q. Sam worked 7 hours for his parents at $8 an hour, and his parents paid him with a $50 bill. Use the symbol ≠ to point out why Sam was upset.

A. **$50 ≠ $56.** Sam worked 7 hours for $8 an hour, so here's how much he earned:

$$7 \times \$8 = \$56$$

He was upset because his parents didn't pay him the correct amount: $50 ≠ $56.

Q. Place the correct symbol (=, >, or <) in the blank: 42 – 19 _____ 5 × 4.

A. **>.** Because 42 – 19 = 23 and 5 × 4 = 20, and 23 is greater than 20, use the symbol that means *is greater than*.

Q. Find an approximate solution to 2,000,398 + 6,001,756.

A. **8,000,000.** The two numbers are both in the millions, so you can use ≈ to round them to the nearest million:

$$2,000,398 + 6,001,756 \approx 2,000,000 + 6,000,000$$

Now it's easy to add 2,000,000 + 6,000,000 = 8,000,000.

11. Place the correct symbol (=, >, or <) in the blanks:

a. 4 + 6 _____ 13

b. 9 × 7 _____ 62

c. 33 – 16 _____ 60 ÷ 3

d. 100 ÷ 5 _____ 83 – 63

Solve It

12. Change the ≠ signs to either > or <:

a. 17 + 14 ≠ 33

b. 144 – 90 ≠ 66

c. 11 × 14 ≠ 98

d. 150 ÷ 6 ≠ 20

Solve It

13. Tim's boss paid him for 40 hours of work last week. Tim accounted for his time by saying that he spent 19 hours with clients, 11 hours driving, and 7 hours doing paperwork. Use ≠ to show why Tim's boss was unhappy with Tim's work.

Solve It

14. Find an approximate solution to 10,002 – 6,007.

Solve It

Special Times: Powers and Square Roots

Raising a number to a *power* is a quick way to multiply a number by itself. For example, 2^5, which you read as *two to the fifth power,* means that you multiply 2 by itself 5 times:

$$2^5 = 2 \times 2 \times 2 \times 2 \times 2 = 32$$

The number 2 is called the *base,* and the number 5 is called the *exponent.*

Powers of ten — that is, powers with 10 in the base — are especially important because the number system is based on them. Fortunately, they're very easy to work with. To raise 10 to the power of any positive whole number, write down the number 1 followed by the number of 0s indicated by the exponent. For example, 10^3 is 1,000.

Here are some important rules for finding powers that contain 0 or 1:

- ✔ Every number raised to the power of 1 equals that number itself.

- ✔ Every number (except 0) raised to the power of 0 is equal to 1. For example, 10^0 is 1 followed by *no* 0s — that is, 1.

✔ The number 0 raised to the power of any number (except 0) equals 0, because no matter how many times you multiply 0 by itself, the result is 0.

Mathematicians have chosen to leave 0^0 undefined — that is, it doesn't equal any number.

✔ The number 1 raised to the power of any number equals 1, because no matter how many times you multiply 1 by itself, the result is 1.

When you multiply any number by itself, the result is a *square number.* So, when you raise any number to the power of 2, you're *squaring* that number. For example, here's 5^2, or *five squared:*

$$5^2 = 5 \times 5 = 25$$

The inverse of squaring a number is called finding the *square root* of a number (inverse operations undo each other — see the earlier section "Switching Things Up with Inverse Operations and the Commutative Property"). When you find the square root of a number, you discover a new number which, when multiplied by itself, equals the number you started with. For example, here's the square root of 25:

$$\sqrt{25} = 5 \text{ (because } 5 \times 5 = 25)$$

Q. What is 3^4?

A. **81.** The expression 3^4 tells you to multiply 3 by itself 4 times:

$$3 \times 3 \times 3 \times 3 = 81$$

Q. What is 10^6?

A. **1,000,000.** Using the power of ten rule, 10^6 is 1 followed by six 0s, so $10^6 =$ 1,000,000.

Q. What is $\sqrt{36}$?

A. **6.** To find $\sqrt{36}$, you want to find a number that, when multiplied by itself, equals 36. You know that $6 \times 6 = 36$, so $\sqrt{36} = 6$.

Q. What is $\sqrt{256}$?

A. **16.** To find $\sqrt{256}$, you want to find a number that, when multiplied by itself, equals 256. Try guessing to narrow down the possibilities. Start by guessing 10:

$$10 \times 10 = 100$$

$256 > 100$, so $\sqrt{256}$ is greater than 10. Guess 20:

$$20 \times 20 = 400$$

$256 < 400$, so $\sqrt{256}$ is between 10 and 20. Guess 15:

$$15 \times 15 = 225$$

$256 > 225$, so $\sqrt{256}$ is between 15 and 20. Guess 16:

$$16 \times 16 = 256$$

This is correct, so $\sqrt{256} = 16$.

15. Find the value of the following powers:

    a. 6^2

    b. 3^5

    c. 2^7

    d. 2^8 (*Hint:* You can make your work easier by using the answer to *c*.)

Solve It

16. Find the value of the following powers:

    a. 10^4

    b. 10^{10}

    c. 10^{15}

    d. 10^1

Solve It

Answers to Problems in Smooth Operators

The following are the answers to the practice questions presented in this chapter.

1 Using inverse operations, write down an alternative form of each equation:

a. 8 + 9 = 17: **17 − 9 = 8**

b. 23 − 13 = 10: **10 + 13 = 23**

c. 15 × 5 = 75: **75 ÷ 5 = 15**

d. 132 ÷ 11 = 12: **12 × 11 = 132**

2 Use the commutative property to write down an alternative form of each equation:

a. 19 + 35 = 54: **35 + 19 = 54**

b. 175 + 88 = 263: **88 + 175 = 263**

c. 22 × 8 = 176: **8 × 22 = 176**

d. 101 × 99 = 9,999: **99 × 101 = 9,999**

3 Use inverse operations and the commutative property to find all three alternative forms for each equation:

a. 7 + 3 = 10: **10 − 3 = 7, 3 + 7 = 10, and 10 − 7 = 3**

b. 12 − 4 = 8: **8 + 4 = 12, 4 + 8 = 12, and 12 − 8 = 4**

c. 6 × 5 = 30: **30 ÷ 5 = 6, 5 × 6 = 30, and 30 ÷ 6 = 5**

d. 18 ÷ 2 = 9: **9 × 2 = 18, 2 × 9 = 18, 18 ÷ 9 = 2**

4 Fill in the blank in each equation:

a. **110.** Rewrite _____ − 74 = 36 as its inverse:

36 + 74 = _____

Therefore, 36 + 74 = 110.

b. **15.** Rewrite _____ × 7 = 105 as its inverse:

105 ÷ 7 = _____

So, 105 ÷ 7 = 15.

c. **87.** Rewrite 45 + _____ = 132 using the commutative property:

_____ + 45 = 132

Now rewrite this equation as its inverse:

132 − 45 = _____

Therefore, 132 − 45 = 87.

d. **203.** Rewrite 273 − _____ = 70 by switching around the two numbers next to the equal sign:

273 − 70 = _____

So, 273 − 70 = 203.

    e. 81. Rewrite $8 \times$ _____ $= 648$ using the commutative property:

    _____ $\times 8 = 648$

    Now rewrite this equation as its inverse:

       $648 \div 8 =$ _____

    So, $648 \div 8 = 81$.

    f. 20. Rewrite $180 \div$ _____ $= 9$ by switching around the two numbers next to the equal sign:

    $180 \div 9 =$ _____ So, $180 \div 9 = 20$.

5 **58.** First, do the multiplication inside the parentheses:

      $(8 \times 6) + 10 = 48 + 10$

Now add: $48 + 10 = 58$.

6 **123.** First, do the subtraction inside the parentheses:

      $123 \div (145 - 144) = 123 \div 1$

Now simply divide $123 \div 1 = 123$.

7 Solve the following two problems:

a. $(40 \div 2) + 6 = 20 + 6 = $ **26**

b. $40 \div (2 + 6) = 40 \div 8 = $ **5**

Yes, the placement of parentheses changes the result.

8 Solve the following two problems:

a. $(16 + 24) + 19 = 40 + 19 = $ **59**

b. $16 + (24 + 19) = 16 + 43 = $ **59**

No, because of the associative property of addition, the placement of parentheses doesn't change the result.

9 Solve the following two problems:

  a. $(18 \times 25) \times 4 = 450 \times 4 = $ **1,800**

  b. $18 \times (25 \times 4) = 18 \times 100 = $ **1,800**

No, because of the associative property of multiplication, the placement of parentheses doesn't change the result.

10 $93,769 \times 2 \times 5 = $ **937,690.** The problem is easiest to solve by placing parentheses around 2×5:

    $93,769 \times (2 \times 5) = 93,769 \times 10 = 937,690$

11 Place the correct symbol ($=$, $>$, or $<$) in the blanks:

a. $4 + 6 = 10$, and $10 < 13$

b. $9 \times 7 = 63$, and $63 > 62$

c. $33 - 16 = 17$ and $60 \div 3 = 20$, so $17 < 20$.

d. $100 \div 5 = 20$ and $83 - 63 = 20$, so $20 = 20$.

12 Change the ≠ signs to either > or <:

a. 17 + 14 = 31, and 31 < 33

b. 144 – 90 = 54, and 54 < 66

c. 11 × 14 = 154, and 154 > 98

d. 150 ÷ 6 = 25, and 25 > 20

13 19 + 11 + 7 = **37 ≠ 40**

14 10,002 – 6,007 ≈ **4,000**

15 Find the value of the following powers:

a. $6^2 = 6 \times 6 =$ **36.**

b. $3^5 = 3 \times 3 \times 3 \times 3 \times 3 =$ **243.**

c. $2^7 = 2 \times 2 \times 2 \times 2 \times 2 \times 2 \times 2 =$ **128.**

d. $2^8 = 2 \times 2 \times 2 \times 2 \times 2 \times 2 \times 2 \times 2 =$ **256.** You already know from part c that $2^7 = 128$, so multiply this number by 2 to get your answer: 128 × 2 = 256.

16 Find the value of the following powers:

a. $10^4 =$ **10,000.** Write 1 followed by four 0s.

b. $10^{10} =$ **10,000,000,000.** Write 1 followed by ten 0s.

c. $10^{15} =$ **1,000,000,000,000,000.** Write 1 followed by fifteen 0s.

d. $10^1 =$ **10.** Any number raised to the power of 1 is that number.

Chapter 3

Getting Down with Negative Numbers

In This Chapter

- Understanding negative numbers
- Finding the absolute value of a number
- Adding and subtracting negative numbers
- Multiplying and dividing negative numbers

Negative numbers, which are commonly used to represent debt and really cold temperatures, represent amounts less than zero. Such numbers arise when you subtract a large number from a smaller one. In this chapter, you apply the Big Four operations (adding, subtracting, multiplying, and dividing) to negative numbers.

Understanding Where Negative Numbers Come From

When you first discovered subtraction, you were probably told that you can't take a small number minus a greater number. For example, if you start with four marbles, you can't subtract six because you can't take away more marbles than you have. This rule is true for marbles, but in other situations, you *can* subtract a big number from a small one. For example, if you have \$4 and you buy something that costs \$6, you end up with less than \$0 dollars — that is, –\$2, which means a *debt* of \$2.

A number with a minus sign in front of it, like –2, is called a *negative number.* You call the number –2 either *negative two* or *minus two.* Negative numbers appear on the number line to the left of 0, as shown in Figure 3-1.

Figure 3-1: Negative numbers on the number line.

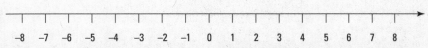

Subtracting a large number from a small number makes sense on the number line. Just use the rule for subtraction that I show you in Chapter 2: Start at the first number and count left the second number of places.

When you don't have a number line to work with, here's a simple rule for subtracting a small number minus a big number: Switch the two numbers around and take the big number minus the small one; then attach a negative sign to the result.

Q. Use the number line to subtract 5 – 8.

A. **–3.** On the number line, 5 – 8 means *start at 5, left 8*.

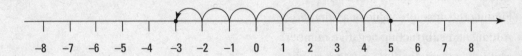

Q. What is 11 – 19?

A. **–8.** Because 11 is less than 19, subtract 19 – 11, which equals 8, and attach a minus sign to the result. Therefore, 11 – 19 = –8.

1. Using the number line, subtract the following numbers:

a. 1 – 4 = _____

b. 3 – 7 = _____

c. 6 – 8 = _____

d. 7 – 14 = _____

Solve It

2. Find the answers to the following subtraction problems:

a. 15 – 22 = _____

b. 27 – 41 = _____

c. 89 – 133 = _____

d. 1,000 – 1,234 = _____

Solve It

Sign-Switching: Understanding Negation and Absolute Value

When you attach a minus sign to any number, you *negate* that number. Negating a number means changing its sign to the opposite sign, so

✔ Attaching a minus sign to a positive number makes it negative.

✔ Attaching a minus sign to a negative number makes it positive. The two *adjacent* (side-by-side) minus signs cancel each other out.

✔ Attaching a minus sign to 0 doesn't change its value, so –0 = 0.

In contrast to negation, placing two bars around a number gives you the absolute value of that number. *Absolute value* is the number's distance from 0 on the number line — that is, it's the positive value of a number, regardless of whether you started out with a negative or positive number:

✔ The absolute value of a positive number is the same number.

✔ The absolute value of a negative number makes it a positive number.

✔ Placing absolute value bars around 0 doesn't change its value, so |0| = 0.

✔ Placing a minus sign outside absolute value bars gives you a negative result — for example, –|6| = –6, and –|–6| = –6.

Q. Negate the number 7.

A. **–7.** Negate 7 by attaching a negative sign to it: –7.

Q. Find the negation of –3.

A. **3.** The negation of –3 is – (–3). The two adjacent minus signs cancel out, which gives you 3.

Q. What's the negation of 7 – 12?

A. **5.** First do the subtraction, which tells you 7 – 12 = –5. To find the negation of –5, attach a minus sign to the answer: –(–5). The two *adjacent* minus signs cancel out, which gives you 5.

Q. What number does |9| equal?

A. **9.** The number 9 is already positive, so the absolute value of 9 is also 9.

Q. What number does |–17| equal?

A. **17.** Because –17 is negative, the absolute value of –17 is 17.

Q. Solve this absolute value problem: –|9 – 13| = ?

A. **–4.** Do the subtraction first: 9 – 13 = –4, which is negative, so the absolute value of –4 is 4. But the minus sign on the left (outside the absolute value bars in the original expression) negates this result, so the answer is –4.

3. Negate each of the following numbers and expressions by attaching a minus sign and then canceling out minus signs when possible:

a. 6

b. −29

c. 0

d. 10 + 4

e. 15 − 7

f. 9 − 10

Solve It

4. Solve the following absolute value problems:

a. |7| = ?

b. |−11| = ?

c. |3 + 15| = ?

d. −|10 − 1| = ?

e. |1 − 10| = ?

f. |0| = ?

Solve It

Adding with Negative Numbers

When you understand what negative numbers mean, you can add them just like the positive numbers you're used to. The number line can help make sense of this. You can turn every problem into a sequence of ups and downs, as I show you in Chapter 1. When you're adding on the number line, starting with a negative number isn't much different from starting with a positive number.

Adding a negative number is the same as subtracting a positive number — that is, go *down* (to the left) on the number line. This rule works regardless of whether you start with a positive or negative number.

After you understand how to add negative numbers on the number line, you're ready to work without the number line. This becomes important when numbers get too large to fit on the number line. Here are some tricks:

- ✔ **Adding a negative number plus a positive number:** Switch around the two numbers (and their signs), turning the problem into subtraction.

- ✔ **Adding a positive number plus a negative number:** Drop the plus sign, turning the problem into subtraction.

- ✔ **Adding two negative numbers:** Drop both minus signs and add the numbers as if they were both positive; then attach a minus sign to the result.

Q. Use the number line to add –3 + 5.

A. **2.** On the number line, –3 + 5 means *start at –3, up 5,* which brings you to 2:

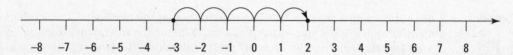

Q. Use the number line to add 6 + –2.

A. **4.** On the number line, 6 + –2 means *start at 6, down 2,* which brings you to 4:

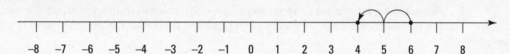

Q. Use the number line to add –3 + –4.

A. **–7.** On the number line, –3 + –4 means *start at –3, down 4,* which brings you to –7:

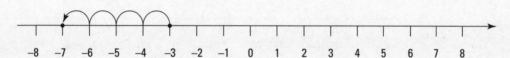

Q. Add –23 + 39.

A. **16.** Switch around the two numbers with the signs attached:

–23 + 39 = + 39 – 23

Now you can drop the plus sign and use the minus sign for subtraction:

39 – 23 = 16

---

5. Use the number line to solve the following addition problems:

a. –5 + 6

b. –1 + –7

c. 4 + –6

d. –3 + 9

e. 2 + –1

f. –4 + –4

Solve It

6. Solve the following addition problems without using the number line:

a. –17 + 35

b. 29 + –38

c. –61 + –18

d. 70 + –63

e. –112 + 84

f. –215 + –322

Solve It

Subtracting with Negative Numbers

Subtracting a negative number is the same as adding a positive number — that is, go up on the number line. This rule works regardless of whether you start with a positive number or a negative number.

When subtracting a negative number, remember that the two back-to-back minus signs cancel each other out, leaving you with a plus sign. (Kind of like when you insist you *can't not* laugh at your friends, because they're really pretty ridiculous; the two negatives mean you *have to laugh,* which is a positive statement.)

Note: Math books often put parentheses around the negative number you're subtracting so the signs don't run together, so 3 – –5 is the same as 3 – (–5).

When taking a negative number minus a positive number, drop both minus signs and add the two numbers as if they were both positive; then attach a minus sign to the result.

Q. Use the number line to subtract –1 – 4.

A. –5. On the number line, –1 – 4 means start at –1, down 4, which brings you to –5:

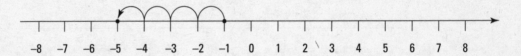

7. Use the number line to solve the following subtraction problems:

   a. –3 – 4

   b. 5 – (–3)

   c. –1 – (–8)

   d. –2 – 4

   e. –4 – 2

   f. –6 – (–10)

Solve It

8. Solve the following subtraction problems without using the number line:

   a. 17 – (–26)

   b. –21 – 45

   c. –42 – (–88)

   d. –67 – 91

   e. 75 – (–49)

   f. –150 – (–79)

Solve It

Knowing Signs of the Times (And Division) for Negative Numbers

Multiplying and dividing negative numbers is basically the same as it is with positive numbers. The presence of one or more minus signs (–) doesn't change the numerical part of the answer. The only question is whether the sign of the answer is positive or negative.

Remember the following when you multiply or divide negative numbers:

✔ If the two numbers have the *same sign,* the result is always positive.

✔ If the two numbers have *opposite signs,* the result is always negative.

Q. Solve the following four multiplication problems:

a. $5 \times 6 =$ _____

b. $-5 \times 6 =$ _____

c. $5 \times -6 =$ _____

d. $-5 \times -6 =$ _____

A. As you can see from this example, the numerical part of the answer (30) doesn't change. Only the sign of the answer changes, depending on the signs of the two numbers in the problem.

a. $5 \times 6 = 30$

b. $-5 \times 6 = -30$

c. $5 \times -6 = -30$

d. $-5 \times -6 = 30$

Q. Solve the following four division problems:

a. $18 \div 3 =$ _____

b. $-18 \div 3 =$ _____

c. $18 \div -3 =$ _____

d. $-18 \div -3 =$ _____

A. The numerical part of the answer (6) doesn't change. Only the sign of the answer changes, depending on the signs of the two numbers in the problem.

a. $18 \div 3 = 6$

b. $-18 \div 3 = -6$

c. $18 \div -3 = -6$

d. $-18 \div -3 = 6$

Q. What is -84×21?

A. **-1,764.** First, drop the signs and multiply:

$$84 \times 21 = 1,764$$

The numbers -84 and 21 have different signs, so the answer is negative: $-1,764$.

Q. What is $-580 \div -20$?

A. **29.** Drop the signs and divide (you can use long division, as I show you in Chapter 1):

$$580 \div 20 = 29$$

The numbers -580 and -20 have the same signs, so the answer is positive: 29.

9. Solve the following multiplication problems:

a. $7 \times 11 =$ _____

b. $-7 \times 11 =$ _____

c. $7 \times -11 =$ _____

d. $-7 \times -11 =$ _____

Solve It

10. Solve the following division problems:

a. $32 \div -8 =$ _____

b. $-32 \div -8 =$ _____

c. $-32 \div 8 =$ _____

d. $32 \div 8 =$ _____

Solve It

11. What is -65×23?

Solve It

12. Find -143×-77.

Solve It

13. Calculate $216 \div -9$.

Solve It

14. What is $-3,375 \div -25$?

Solve It

Answers to Problems in Getting Down with Negative Numbers

The following are the answers to the practice questions presented in this chapter.

1. Using the number line, subtract the following numbers:

   a. $1 - 4 = -3$. Start at 1, down 4.

   b. $3 - 7 = -4$. Start at 3, down 7.

   c. $6 - 8 = -2$. Start at 6, down 8.

   d. $7 - 14 = -7$. Start at 7, down 14.

2. Find the answers to the following subtraction problems:

   a. $15 - 22 = $ **–7.** Fifteen is less than 22, so subtract $22 - 15 = 7$ and attach a minus sign to the result: –7.

   b. $27 - 41 = $ **–14.** Twenty-seven is less than 41, so subtract $41 - 27 = 14$ and attach a minus sign to the result: –14.

   c. $89 - 133 = $ **–44.** Eighty-nine is less than 133, so subtract $133 - 89 = 44$ and attach a minus sign to the result: –44.

   d. $1,000 - 1,234 = $ **–234.** One thousand is less than 1,234, so subtract $1,234 - 1,000 = 234$ and attach a minus sign to the result: –234.

3. Negate each of the following numbers and expressions by attaching a minus sign and then canceling out minus signs when possible:

   a. **–6.** To negate 6, attach a minus sign: –6.

   b. **29.** To negate –29, attach a minus sign: $-(29)$. Now cancel adjacent minus signs: $-(-29) = 29$.

   c. **0.** Zero is its own negation.

   d. **–14.** Add first: $10 + 4 = 14$, and the negation of 14 is –14.

   e. **–8.** Subtract first: $15 - 7 = 8$, and the negation of 8 is –8.

   f. **1.** Begin by subtracting: $9 - 10 = -1$, and the negation of –1 is 1.

4. Solve the following absolute value problems:

   a. $|7| = $ **7.** Seven is already positive, so the absolute value of 7 is also 7.

   b. $|-11| = $ **11.** The number –11 is negative, so the absolute value of –11 is 11.

   c. $|3 + 15| = $ **18.** First, do the addition inside the absolute value bars: $3 + 15 = 18$, which is positive. The absolute value of 18 is 18.

   d. $-|10 - 1| = $ **–9.** First do the subtraction: $10 - 1 = 9$, which is positive. The absolute value of 9 is 9. You have a negative sign outside the absolute value bars, so negate your answer to get –9.

   e. $|1 - 10| = $ **9.** Begin by subtracting: $1 - 10 = -9$, which is negative. The absolute value of –9 is 9.

   f. $|0| = $ **0.** The absolute value of 0 is 0.

5 Use the number line to solve the following addition problems:

a. $-5 + 6 = $ **1.** Start at –5, go up 6.

b. $-1 + -7 = $ **–8.** Start at –1, go down 7.

c. $4 + -6 = $ **–2.** Start at 4, go down 6.

d. $-3 + 9 = $ **6.** Start at –3, go up 9.

e. $2 + -1 = $ **1.** Start at 2, go down 1.

f. $-4 + -4 = $ **–8.** Start at –4, go down 4.

6 Solve the following addition problems without using the number line:

a. $-17 + 35 = $ **18.** Switch around the numbers (with their signs) to turn the problem into subtraction:

$-17 + 35 = 35 - 17 = 18$

b. $29 + -38 = $ **–9.** Drop the plus sign to turn the problem into subtraction:

$29 + -38 = 29 - 38 = -9$

c. $-61 + -18 = $ **–79.** Drop the signs, add the numbers, and negate the result:

$61 + 18 = 79$, so $-61 + -18 = -79$.

d. $70 + -63 = $ **7.** Turn the problem into subtraction:

$70 + -63 = 70 - 63 = 7$

e. $-112 + 84 = $ **–28.** Turn the problem into subtraction:

$-112 + 84 = 84 - 112 = -28$

f. $-215 + -322 = $ **–537.** Drop the signs, add the numbers, and negate the result:

$215 + 322 = 537$, so $-215 + -322 = -537$

7 Use the number line to solve the following subtraction problems:

a. $-3 - 4 = $ **–7.** Start at –3, down 4.

b. $5 - (-3) = $ **8.** Start at 5, up 3.

c. $-1 - (-8) = $ **7.** Start at –1, up 8.

d. $-2 - 4 = $ **–6.** Start at –2, down 4.

e. $-4 - 2 = $ **–6.** Start at –4, down 2.

f. $-6 - (-10) = $ **4.** Start at –6, up 10.

8 Solve the following subtraction problems without using the number line:

a. $17 - (-26) = $ **43.** Cancel the adjacent minus signs to turn the problem into addition:

$17 - (-26) = 17 + 26 = 43$

b. $-21 - 45 = $ **–66.** Drop the signs, add the numbers, and negate the result:

$21 + 45 = 66$, so $-21 - 45 = -66$

c. $-42 - (-88) = $ **46.** Cancel the adjacent minus signs to turn the problem into addition:

$-42 - (-88) = -42 + 88$

Now switch around the numbers (with their signs) to turn the problem back into subtraction:

$88 - 42 = 46$

d. $-67 - 91 = $ **−158.** Drop the signs, add the numbers, and negate the result:

$67 + 91 = 158$, so $-67 - 91 = -158$

e. $75 - (-49) = $ **124.** Cancel the adjacent minus signs to turn the problem into addition:

$75 - (-49) = 75 + 49 = 124$

f. $-150 - (-79) = $ **−71.** Cancel the adjacent minus signs to turn the problem into addition:

$-150 - (-79) = -150 + 79$

Now switch around the numbers (with their signs) to turn the problem back into subtraction:

$79 - 150 = -71$

9 Solve the following multiplication problems:

a. $7 \times 11 = $ **77**

b. $-7 \times 11 = $ **−77**

c. $7 \times -11 = $ **−77**

d. $-7 \times -11 = $ **77**

10 Solve the following division problems:

a. $32 \div -8 = $ **−4**

b. $-32 \div -8 = $ **4**

c. $-32 \div 8 = $ **−4**

d. $32 \div 8 = $ **4**

11 $-65 \times 23 = $ **−1,495.** First, drop the signs and multiply:

$65 \times 23 = 1,495$

The numbers −65 and 23 have different signs, so the answer is negative: −1,495.

12 $-143 \times -77 = $ **11,011.** Drop the signs and multiply:

$143 \times 77 = 11,011$

The numbers −143 and −77 have the same sign, so the answer is positive: 11,011.

13 $216 \div -9 = $ **−24.** Drop the signs and divide (use long division, as I show you in Chapter 1):

$216 \div 9 = 24$

The numbers 216 and −9 have different signs, so the answer is negative: −24.

14 $-3,375 \div -25 = $ **135.** First, drop the signs and divide:

$3,375 \div 25 = 135$

The numbers −3,375 and −25 have the same sign, so the answer is positive: 135.

Chapter 4

It's Just an Expression

An arithmetic expression is any string of numbers and operators that can be calculated. In some cases, the calculation is easy. For example, you can calculate 2 + 2 in your head to come up with the answer 4. As expressions become longer, however, the calculation becomes more difficult. You may have to spend more time with the expression $2 \times 6 + 23 - 10 + 13$ to find the correct answer of 38.

The word *evaluate* comes from the word *value*. When you *evaluate* an expression, you turn it from a string of mathematical symbols into a single value — that is, you turn it into one number. But as expressions get more complicated, the potential for confusion arises. For example, think about the expression $3 + 2 \times 4$. If you add the first two numbers and then multiply, your answer is 20. But if you multiply the last two numbers and then add, your answer is 11.

To solve this problem, mathematicians have agreed on an *order of operations* (sometimes called *order of precedence*): a set of rules for deciding how to evaluate an arithmetic expression no matter how complex it gets. In this chapter, I introduce you to the order of operations through a series of exercises that introduce the basic concepts one at a time. When you finish this chapter, you should be able to evaluate just about any expression your teacher can give you!

Evaluating Expressions with Addition and Subtraction

When an expression has *only* addition and subtraction, in any combination, it's easy to evaluate: Just start with the first two numbers and continue from left to right. Even when an expression includes negative numbers, the same procedure applies (just make sure you use the correct rule for adding or subtracting negative numbers, as I discuss in Chapter 3).

Q. Find $7 + -3 - 6 - (-10)$.

A. **8.** Start at the left with the first two numbers, $7 + -3 = 4$:

$$\underline{7 + -3} - 6 - (-10) = \underline{4} - 6 - (-10)$$

Continue with the next two numbers, $4 - 6 = -2$:

$$\underline{4 - 6} - (-10) = \underline{-2} - (-10)$$

Finish up with the last two numbers, remembering that subtracting a negative number is the same thing as adding a positive number:

$$-2 - (-10) = -2 + 10 = 8$$

---

1. What's $9 - 3 + 8 - 7$?

Solve It

2. Evaluate $11 - 5 - 2 + 6 - 12$.

Solve It

---

3. Find $17 - 11 - (-4) + -10 - 8$.

Solve It

4. $-7 + -3 - (-11) + 8 - 10 + -20 = ?$

Solve It

Evaluating Expressions with Multiplication and Division

Unless you've been using your free time to practice your Persian, Arabic, or Hebrew reading skills, you're probably used to reading from left to right. Luckily, that's the direction of choice for multiplication and division problems, too. When an expression has *only* multiplication and division, in any combination, you should have no trouble evaluating it: Just start with the first two numbers and continue from left to right.

Q. What is $15 \div 5 \times 8 \div 6$?

A. **4.** Start at the left with the first two numbers, $15 \div 5 = 3$:

$$\underline{15 \div 5} \times 8 \div 6 = \underline{3} \times 8 \div 6$$

Continue with the next two numbers, $3 \times 8 = 24$:

$$\underline{3 \times 8} \div 6 = \underline{24} \div 6$$

Finish up with the last two numbers:

$$24 \div 6 = 4$$

Q. Evaluate $-10 \times 2 \times -3 \div -4$.

A. **–15.** The same procedure applies when you have negative numbers (just make sure you use the correct rule for multiplying or dividing by negative numbers, as I explain in Chapter 3). Start at the left with the first two numbers, $-10 \times 2 = -20$:

$$\underline{-10 \times 2} \times -3 \div -4 = \underline{-20} \times -3 \div -4$$

Continue with the next two numbers, $-20 \times -3 = 60$:

$$\underline{-20 \times -3} \div -4 = \underline{60} \div -4$$

Finish up with the last two numbers:

$$60 \div -4 = -15$$

5. Find $18 \div 6 \times 10 \div 6$.

Solve It

6. Evaluate $20 \div 4 \times 8 \div 5 \div -2$.

Solve It

7. $12 \div -3 \times -9 \div 6 \times -7 = ?$

Solve It

8. Solve $-90 \div 9 \times -8 \div -10 \div 4 \times 15$.

Solve It

Making Sense of Mixed-Operator Expressions

Things get a little complicated in this section, but you can handle it. A *mixed-operator* expression contains at least one addition or subtraction sign and at least one multiplication or division sign. To evaluate mixed-operator expressions, follow a couple of simple steps:

1. **Evaluate all multiplication and division from left to right.**

   Begin evaluating a mixed-operator expression by underlining all the multiplication or division in the problem.

2. **Evaluate addition and subtraction from left to right.**

Q. What's $-15 \times 3 \div -5 - (-3) \times -4$?

A. **–3.** Start by underlining all the multiplication and division in the problem; then evaluate all multiplication and division from left to right:

$$\underline{-15 \times 3} \div -5 - (-3) \times -4$$
$$= -45 \div -5 - (-3) \times -4$$

$$= 9 - (-3) \times -4$$
$$= 9 - 12$$

Finish up by evaluating the addition and subtraction from left to right:

$$= -3$$

9. Evaluate $8 - 3 \times 4 \div 6 + 1$.

`Solve It`

10. Find $10 \times 5 - (-3) \times 8 \div -2$.

`Solve It`

11. $-19 - 7 \times 3 + -20 \div 4 - 8 = ?$

`Solve It`

12. What's $60 \div -10 - (-2) + 11 \times 8 \div 2$?

`Solve It`

Handling Powers Responsibly

You may have heard that power corrupts, but rest assured that when mathematicians deal with powers, the order of operations usually keeps them in line. When an expression contains one or more powers, evaluate all powers from left to right before moving on to the Big Four operators. Here's the breakdown:

1. **Evaluate all powers from left to right.**

   In Chapter 2, I show you that raising a number to a power simply means multiplying the number by itself that many times. For example, $2^3 = 2 \times 2 \times 2 = 8$. Remember that anything raised to the 0 power equals 1.

2. **Evaluate all multiplication and division from left to right.**

3. **Evaluate addition and subtraction from left to right.**

If you compare this numbered list with the one in the preceding section, you'll notice the only difference is that I've now inserted a new rule at the top.

Q. Evaluate $7 - 4^2 \div 2^4 + 9 \times 2^3$.

A. **78.** Evaluate all powers from left to right, starting with $4^2 = 4 \times 4 = 16$:

$$= 7 - 16 \div 2^4 + 9 \times 2^3$$

Move on to evaluate the remaining two powers:

$$= 7 - 16 \div 16 + 9 \times 8$$

Next, evaluate all multiplication and division from left to right:

$$= 7 - 1 + 9 \times 8$$

$$= 7 - 1 + 72$$

Finish up by evaluating the addition and subtraction from left to right:

$$= 6 + 72$$

$$= 78$$

---

13. Evaluate $3^2 - 2^3 \div 2^2$.

Solve It

14. Find $5^2 - 4^2 - (-7) \times 2^2$.

Solve It

15. $70^1 - 3^4 \div -9 \times -7 + 123^0 = ?$

Solve It

16. What's $11^2 - 2^7 + 3^5 \div 3^3$?

Solve It

Prioritizing Parentheses

Did you ever go to the post office and send a package high-priority so that it'd arrive as soon as possible? Parentheses work just like that. Parentheses — ( ) — allow you to indicate that a piece of an expression is high-priority — that is, it has to be evaluated before the rest of the expression.

When an expression includes parentheses with only Big Four operators, just do the following:

1. **Evaluate the contents of the parentheses.**

2. **Evaluate Big Four operators (as I show you earlier in "Making Sense of Mixed-Operator Expressions").**

When an expression has more than one set of parentheses, don't panic. Start by evaluating the contents of the first set and move left to right. Piece of cake!

Q. Evaluate $(6 - 2) + (15 ÷ 3)$.

A. **9.** Start by evaluating the contents of the first set of parentheses:

$$(6 - 2) + (15 ÷ 3) = 4 + (15 ÷ 3)$$

Move on to the next set of parentheses:

$$= 4 + 5$$

To finish up, evaluate the addition:

$$= 9$$

Q. Evaluate $(6 + 1) × (5 - (-14) ÷ -7)$.

A. **21.** When a set of parentheses includes a mixed-operator expression, evaluate *everything* inside the parentheses according to the order of operations (see "Making Sense of Mixed-Operator Expressions"). Begin by evaluating the contents of the first set of parentheses: $6 + 1 = 7$:

$$(6 + 1) × (5 - (-14) ÷ -7)$$
$$= 7 × (5 - (-14) ÷ -7)$$

Move to the next set of parentheses. This set contains a mixed-operator expression, so start with the division: $-14 ÷ -7 = 2$:

$$= 7 × (5 - 2)$$

Complete the contents of the parentheses by evaluating the subtraction: $5 - 2 = 3$:

$$= 7 × 3$$

Finish up by evaluating the multiplication: $7 × 3 = 21$.

17. Evaluate $4 \times (3 + 4) - (16 \div 2)$.

Solve It

18. What's $(5 + -8 \div 2) + (3 \times 6)$?

Solve It

19. Find $(4 + 12 \div 6 \times 7) - (3 + 8)$.

Solve It

20. $(2 \times -5) - (10 - 7) \times (13 + -8) = ?$

Solve It

Pulling Apart Parentheses and Powers

When an expression has parentheses and powers, evaluate it in the following order:

1. **Contents of parentheses**

An expression in an *exponent* (a small, raised number indicating a power) groups that expression like parentheses do. Evaluate any superscripted expression down to a single number before evaluating the power. In other words, to find 5^{3-1}, you can pretend $3 - 1$ is in parentheses, making the problem $5^{(3-1)} = 5^2 = 25$.

A few other symbols that you may be familiar with also group expressions together just like parentheses. These include the square root symbol and absolute value bars, which I introduce in Chapter 2 and Chapter 3, respectively.

2. **Powers from left to right**

3. **Multiplication and division from left to right**

4. **Addition and subtraction from left to right**

Compare this numbered list with the one in the previous section, "Handling Powers Responsibly." The only real difference is that I've now inserted a new rule at the top.

EXAMPLE

Q. Evaluate $(8 + 6^2) \div (2^3 - 4)$.

A. **11.** Begin by evaluating the contents of the first set of parentheses. Inside this set, evaluate the power first and do the addition next:

$$(8 + 6^2) \div (2^3 - 4)$$

$$= (8 + 36) \div (2^3 - 4)$$

$$= 44 \div (2^3 - 4)$$

Move to the next set of parentheses, evaluating the power first and then the subtraction:

$$= 44 \div (8 - 4) = 44 \div 4$$

Finish up by evaluating the division: $44 \div 4 = 11$.

Q. Find the value of $-1 + (-20 + 3^3)^2$.

A. **48.** When the entire contents of a set of parentheses is raised to a power, evaluate what's inside the parentheses before evaluating the power. *Inside* this set, evaluate the power first and the addition next:

$$-1 + (-20 + 3^3)^2 = -1 + (-20 + 27)^2 = -1 + 7^2$$

Next, evaluate the power $7^2 = 7 \times 7 = 49$:

$$= -1 + 49$$

Finish up by evaluating the addition: $-1 + 49 = 48$

21. Find $(6^2 - 12) \div (16 \div 2^3)$.

Solve It

22. Evaluate $-10 - (2 + 3^2 \times -4)$.

Solve It

23. $7^2 - (3 + 3^2 \div -9)^5 = ?$

Solve It

24. What is $(10 - 1^{14} \times 8)^{4+4+5}$

Solve It

Figuring Out Nested Parentheses

Have you ever seen nested wooden dolls? (They originated in Russia, where their ultra-cool name is *matryoshka*.) This curiosity appears to be a single carved piece of wood in the shape of a doll. But when you open it up, you find a smaller doll nested inside it. And when you open up the smaller doll, you find an even smaller one hidden inside that one — and so on.

Like these Russian dolls, some arithmetic expressions contain sets of *nested* parentheses — one set of parentheses inside another set. To evaluate a set of nested parentheses, start by evaluating the inner set of parentheses and work your way outward.

Parentheses — ( ) — come in a number of styles, including brackets — [ ] — and braces — { }. These different styles help you keep track of where a statement in parentheses begins and ends. No matter what they look like, to the mathematician these different styles are all parentheses, so they all get treated the same.

Q. Find the value of $\{3 \times [10 \div (6 - 4)]\} + 2$.

A. **17.** Begin by evaluating what's inside the innermost set of parentheses: $6 - 4 = 2$:

$$\{3 \times [10 \div (6 - 4)]\} + 2 = \{3 \times [10 \div 2]\} + 2$$

The result is an expression with one set of parentheses inside another set, so evaluate what's inside the inner set: $10 \div 2 = 5$:

$$= \{3 \times 5\} + 2$$

Now, evaluate what's inside the final set of parentheses:

$$= 15 + 2$$

Finish up by evaluating the addition: $15 + 2 = 17$.

---

25. Evaluate $7 + \{[(10 - 6) \times 5] + 13\}$.

Solve It

26. Find the value of $[(2 + 3) - (30 \div 6)] + (-1 + 7 \times 6)$.

Solve It

27. $-4 + \{[-9 \times (5 - 8)] \div 3\} = ?$

Solve It

28. Evaluate $\{(4 - 6) \times [18 \div (12 - 3 \times 2)]\} - (-5)$.

Solve It

Bringing It All Together: The Order of Operations

 Throughout this chapter, you work with a variety of rules for deciding how to evaluate arithmetic expressions. These rules all give you a way to decide the order in which an expression gets evaluated. All together, this set of rules is called the *order of operations* (or sometimes, the *order of precedence*). Here's the complete order of operations for arithmetic:

1. **Contents of parentheses from the inside out**

2. **Powers from left to right**

3. **Multiplication and division from left to right**

4. **Addition and subtraction from left to right**

The only difference between this list and the one in "Pulling Apart Parentheses and Powers" is that I've now added a few words to the end of Step 1 to cover nested parentheses (which I discuss in the preceding section).

 Q. Evaluate $[(8 \times 4 + 2^3) \div 10]^{7-5}$.

A. **16.** Start by focusing on the inner set of parentheses, evaluating the power, then the multiplication, and then the addition:

$$[(8 \times 4 + 2^3) \div 10]^{7-5}$$

$$= [(8 \times 4 + 8) \div 10]^{7-5}$$

$$= [(32 + 8) \div 10]^{7-5}$$

$$= [40 \div 10]^{7-5}$$

Next, evaluate what's inside the parentheses and the expression that makes up the exponent:

$$= 4^{7-5} = 4^2$$

Finish by evaluating the remaining power: $4^2 = 16$.

29. Evaluate $1 + [(2^3 - 4) + (10 \div 2)^2]$.

Solve It

30. $(-7 \times -2 + 6^2 \div 4)^{9 \times 2 - 17}$

Solve It

31. What is $\{6^2 - [12 \div (-13 + 14)^2] \times 2\}^2$?

Solve It

32. Find the value of $[(123 - 11^2)^4 - (6^2 \div 2^{20 - 3 \times 6})]^2$.

Solve It

Solutions to It's Just an Expression

The following are the answers to the practice questions presented earlier in this chapter.

1 $9 - 3 + 8 - 7 = $ **7.** Add and subtract from left to right:

$$9 - 3 + 8 - 7$$
$$= 6 + 8 - 7$$
$$= 14 - 7 = 7$$

2 $11 - 5 - 2 + 6 - 12 = $ **−2.**

$$11 - 5 - 2 + 6 - 12$$
$$= 6 - 2 + 6 - 12$$
$$= 4 + 6 - 12$$
$$= 10 - 12 = -2$$

3 $17 - 11 - (-4) + -10 - 8 = $ **−8.**

$$17 - 11 - (-4) + -10 - 8$$
$$= 6 - (-4) + -10 - 8$$
$$= 10 + -10 - 8$$
$$= 0 - 8 = -8$$

4 $-7 + -3 - (-11) + 8 - 10 + -20 = $ **−21.**

$$-7 + -3 - (-11) + 8 - 10 + -20$$
$$= -10 - (-11) + 8 - 10 + -20$$
$$= 1 + 8 - 10 + -20$$
$$= 9 - 10 + -20$$
$$= -1 + -20 = -21$$

5 $18 \div 6 \times 10 \div 6 = $ **5.** Divide and multiply from left to right:

$$18 \div 6 \times 10 \div 6$$
$$= 3 \times 10 \div 6$$
$$= 30 \div 6 = 5$$

6 $20 \div 4 \times 8 \div 5 \div -2 = $ **−4.**

$$20 \div 4 \times 8 \div 5 \div -2$$
$$= 5 \times 8 \div 5 \div -2$$
$$= 40 \div 5 \div -2$$
$$= 8 \div -2 = -4$$

7 $12 \div -3 \times -9 \div 6 \times -7 =$ **−42.**

$$12 \div -3 \times -9 \div 6 \times -7$$

$$= -4 \times -9 \div 6 \times -7$$

$$= 36 \div 6 \times -7$$

$$= 6 \times -7 = -42$$

8 $-90 \div 9 \times -8 \div -10 \div 4 \times -15 =$ **30.**

$$-90 \div 9 \times -8 \div -10 \div 4 \times -15$$

$$= -10 \times -8 \div -10 \div 4 \times -15$$

$$= 80 \div -10 \div 4 \times -15$$

$$= -8 \div 4 \times -15$$

$$= -2 \times -15$$

$$= 30$$

9 $8 - 3 \times 4 \div 6 \div 1 =$ **7.** Start by underlining and evaluating all multiplication and division from left to right:

$$8 - \underline{3 \times 4} \div 6 \div 1$$

$$= 8 - \underline{12 \div 6} + 1$$

$$= 8 - \underline{2} + 1$$

Now evaluate all addition and subtraction from left to right:

$$= 6 + 1 = 7$$

10 $10 \times 5 - (-3) \times 8 \div -2 =$ **38.** Start by underlining and evaluating all multiplication and division from left to right:

$$\underline{10 \times 5} - (-3) \times 8 \div -2$$

$$= 50 - \underline{(-3) \times 8} \div -2$$

$$= 50 - \underline{(-24) \div -2}$$

$$= 50 - 12$$

Now evaluate the subtraction:

$$= 38$$

11 $-19 - 7 \times 3 + -20 \div 4 - 8 =$ **−53.** Start by underlining and evaluating all multiplication and division from left to right:

$$-19 - \underline{7 \times 3} + \underline{-20 \div 4} - 8$$

$$= -19 - 21 + \underline{-20 \div 4} - 8$$

$$= -19 - 21 + -5 - 8$$

Then evaluate all addition and subtraction from left to right:

$$= -40 + -5 - 8 = -45 - 8 = -53$$

12 $60 \div -10 - (-2) + 11 \times 8 \div 2 =$ **40.** Start by underlining and evaluating all multiplication and division from left to right:

$\underline{60 \div -10} - (-2) + \underline{11 \times 8} \div 2$

$= -6 - (-2) + \underline{11 \times 8} \div 2$

$= -6 - (-2) + \underline{88 \div 2}$

$= -6 - (-2) + 44$

Now evaluate all addition and subtraction from left to right:

$= -4 + 44 = 40$

13 $3^2 - 2^3 \div 2^2 =$ **7.** First, evaluate all powers:

$3^2 - 2^3 \div 2^2 = 9 - 8 \div 4$

Next, evaluate the division:

$= 9 - 2$

Finally, evaluate the subtraction:

$9 - 2 = 7$

14 $5^2 - 4^2 - (-7) \times 2^2 =$ **37.** Evaluate all powers:

$5^2 - 4^2 - (-7) \times 2^2 = 25 - 16 - (-7) \times 4$

Evaluate the multiplication:

$= 25 - 16 - (-28)$

Finally, evaluate the subtraction from left to right:

$= 9 - (-28) = 37$

15 $70^1 - 3^4 \div -9 \times -7 + 123^0 =$ **8.** Evaluate all powers:

$70^1 - 3^4 \div -9 \times -7 + 123^0 = 70 - 81 \div -9 \times -7 + 1$

Next, evaluate the multiplication and division from left to right:

$= 70 - (-9) \times -7 + 1 = 70 - 63 + 1$

Evaluate the addition and subtraction from left to right:

$= 7 + 1 = 8$

16 $11^2 - 2^7 + 3^5 \div 3^3 =$ **2.** First, evaluate all powers:

$11^2 - 2^7 + 3^5 \div 3^3 = 121 - 128 + 243 \div 27$

Evaluate the division:

$= 121 - 128 + 9$

Finally, evaluate the addition and subtraction from left to right:

$= -7 + 9 = 2$

17 $4 \times (3 + 4) - (16 \div 2) =$ **20.** Start by evaluating what's inside the first set of parentheses:

$$4 \times (3 + 4) - (16 \div 2)$$

$$= 4 \times 7 - (16 \div 2)$$

Next, evaluate the contents of the second set of parentheses:

$$4 \times 7 - 8$$

Evaluate the multiplication and then the subtraction:

$$= 28 - 8 = 20$$

18 $(5 + -8 \div 2) + (3 \times 6) =$ **19.** Inside the first set of parentheses, evaluate the division first and then the addition:

$$(5 + -8 \div 2) + (3 \times 6)$$

$$= (5 + -4) + (3 \times 6)$$

$$= 1 + (3 \times 6)$$

Next, evaluate the contents of the second set of parentheses:

$$= 1 + 18$$

Finish up by evaluating the addition:

$$1 + 18 = 19$$

19 $(4 + 12 \div 6 \times 7) - (3 + 8) =$ **7.** Begin by focusing on the first set of parentheses, handling all multiplication and division from left to right:

$$(4 + 12 \div 6 \times 7) - (3 + 8)$$

$$= (4 + 2 \times 7) - (3 + 8)$$

$$= (4 + 14) - (3 + 8)$$

Now do the addition inside the first set of parentheses:

$$= 18 - (3 + 8)$$

Next, evaluate the contents of the second set of parentheses:

$$= 18 - 11$$

Finish up by evaluating the subtraction:

$$18 - 11 = 7$$

20 $(2 \times -5) - (10 - 7) \times (13 + -8) =$ **−25.** Evaluate the first set of parentheses, then the second, and then the third:

$$(2 \times -5) - (10 - 7) \times (13 + -8)$$

$$= -10 - (10 - 7) \times (13 + -8)$$

$$= -10 - 3 \times (13 + -8)$$

$$= -10 - 3 \times 5$$

Next, do multiplication and then finish up with the subtraction:

$$= -10 - 15 = -25$$

21 $(6^2 - 12) \div (16 \div 2^3) =$ **12.** Focusing on the contents of the first set of parentheses, evaluate the power and then the subtraction:

$$(6^2 - 12) \div (16 \div 2^3)$$

$$= (36 - 12) \div (16 \div 2^3)$$

$$= 24 \div (16 \div 2^3)$$

Next, work inside the second set of parentheses, evaluating the power first and then the division:

$$= 24 \div (16 \div 8) = 24 \div 2$$

Finish by evaluating the division:

$$= 24 \div 2 = 12$$

22 $-10 - (2 + 3^2 \times -4) =$ **24.** Focusing on the contents of the parentheses, evaluate the power first, then the multiplication, and then the addition:

$$-10 - (2 + 3^2 \times -4) = -10 - (2 + 9 \times -4) = -10 - (2 + -36) = -10 - (-34)$$

Finish by evaluating the subtraction:

$$-10 - (-34) = 24$$

23 $7^2 - (3 + 3^2 \div -9)^5 =$ **17.** Focusing *inside* the parentheses, evaluate the power first, then the division, and then the addition:

$$7^2 - (3 + 3^2 \div -9)^5$$

$$= 7^2 - (3 + 9 \div -9)^5$$

$$= 7^2 - (3 + -1)^5$$

$$= 7^2 - 2^5$$

Next, evaluate both powers in order:

$$= 49 - 2^5 = 49 - 32$$

To finish, evaluate the subtraction:

$$49 - 32 = 17$$

24 $(10 - 1^{14} \times 8)^{4 \div 4 + 5} =$ **64.** Focusing inside the first set of parentheses, evaluate the power first, then the multiplication, and then the subtraction:

$$(10 - 1^{14} \times 8)^{4 \div 4 + 5} = (10 - 1 \times 8)^{4 \div 4 + 5} = (10 - 8)^{4 \div 4 + 5} = 2^{4 \div 4 + 5}$$

Next, handle the expression in the exponent, evaluating the division first and then the addition:

$$2^{1 + 5} = 2^6$$

To finish, evaluate the power:

$$2^6 = 64$$

25 $7 + \{[(10 - 6) \times 5] + 13\} = $ **40.** First evaluate the inner set of parentheses:

$7 + \{[(10 - 6) \times 5] + 13\} = 7 + \{[4 \times 5] + 13\}$

Move outward to the next set of parentheses:

$= 7 + \{20 + 13\}$

Next, handle the remaining set of parentheses:

$= 7 + 33$

To finish, evaluate the addition:

$7 + 33 = 40$

26 $[(2 + 3) - (30 \div 6)] + (-1 + 7 \times 6) = $ **41.** Start by focusing on the first set of parentheses. This set contains two inner sets of parentheses, so evaluate these two sets from left to right:

$[(2 + 3) - (30 \div 6)] + (-1 + 7 \times 6)$

$= [(5) - (30 \div 6)] + (-1 + 7 \times 6)$

$= [5 - 5] + (-1 + 7 \times 6)$

Now, the expression has two separate sets of parentheses, so evaluate the first set:

$= 0 + (-1 + 7 \times 6)$

Handle the remaining set of parentheses, evaluating the multiplication first and then the addition:

$= 0 + (-1 + 42) = 0 + 41$

To finish, evaluate the addition:

$0 + 41 = 41$

27 $-4 + \{[-9 \times (5 - 8)] \div 3\} = $ **5.** Start by evaluating the inner set of parentheses:

$-4 + \{[-9 \times (5 - 8)] \div 3\} = -4 + \{[-9 \times -3)] \div 3\}$

Move outward to the next set of parentheses:

$= -4 + [27 \div 3]$

Next, handle the remaining set of parentheses:

$= -4 + 9$

Finally, evaluate the addition:

$-4 + 9 = 5$

28 $\{(4-6) \times [18 \div (12-3 \times 2)]\} - (-5) = $ **–1.** Focus on the inner set of parentheses, $(12-3 \times 2)$. Evaluate the multiplication first and then the subtraction:

$$\{(4-6) \times [18 \div (12-3 \times 2)]\} - (-5)$$

$$= \{(4-6) \times [18 \div (12-6)]\} - (-5)$$

$$= \{(4-6) \times [18 \div 6]\} - (-5)$$

Now the expression is an outer set of parentheses with two inner sets. Evaluate these two inner sets of parentheses from left to right:

$$= \{-2 \times [18 \div 6]\} - (-5) = \{-2 \times 3\} - (-5)$$

Next, evaluate the final set of parentheses:

$$= -6 - (-5)$$

Finish by evaluating the subtraction:

$$-6 - (-5) = -1$$

29 $1 + [(2^3 - 4) + (10 \div 2)^2] = $ **30.** Start by focusing on the first of the two inner sets of parentheses, $(2^3 - 4)$. Evaluate the power first and then the subtraction:

$$1 + [(2^3 - 4) + (10 \div 2)^2] = 1 + [(8 - 4) + (10 \div 2)^2] = 1 + [4 + (10 \div 2)^2]$$

Continue by focusing on the remaining inner set of parentheses:

$$= 1 + [4 + 5^2]$$

Next, evaluate what's inside the last set of parentheses, evaluating the power first and then the addition:

$$= 1 + [4 + 25] = 1 + 29$$

Finish by adding the remaining numbers:

$$1 + 29 = 30$$

30 $(-7 \times -2 + 6^2 \div 4)^{9 \times 2 - 17} = $ **23.** Start with the first set of parentheses. Evaluate the power first, then the multiplication and division from left to right, and then the addition:

$$(-7 \times -2 + 6^2 \div 4)^{9 \times 2 - 17}$$

$$= (-7 \times -2 + 36 \div 4)^{9 \times 2 - 17}$$

$$= (14 + 36 \div 4)^{9 \times 2 - 17}$$

$$= (14 + 9)^{9 \times 2 - 17}$$

$$= 23^{9 \times 2 - 17}$$

Next, work on the exponent, evaluating the multiplication first and then the subtraction:

$$= 23^{18 - 17} = 23^1$$

Finish by evaluating the power:

$$23^1 = 23$$

31 $\{6^2 - [12 \div (-13 + 14)^2] \times 2\}^2 = $ **144.** Start by evaluating the inner set of parentheses $(-13 + 14)$:

$$\{6^2 - [12 \div (-13 + 14)^2] \times 2\}^2$$

$$= \{6^2 - [12 \div 1^2] \times 2\}^2$$

Move outward to the next set of parentheses, $[12 \div 1^2]$, evaluating the power and then the division:

$$= \{6^2 - [12 \div 1] \times 2\}^2$$

$$= \{6^2 - 12 \times 2\}^2$$

Next, work on the remaining set of parentheses, evaluating the power, then the multiplication, and then the subtraction:

$$= \{36 - 12 \times 2\}^2$$

$$= \{36 - 24\}^2$$

$$= 12^2$$

Finish by evaluating the power:

$$12^2 = 144$$

32 $[(123 - 11)^2)^4 - (6^2 \div 2^{20-3\times6})]^2 = $ **49.** Start by working on the exponent, $20 - 3 \times 6$, evaluating the multiplication and then the subtraction:

$$[(123 - 11)^2)^4 - (6^2 \div 2^{20-3\times6})]^2$$

$$= [(123 - 11)^2)^4 - (6^2 \div 2^{20-18})]^2$$

$$= [(123 - 11)^2)^4 - (6^2 \div 2^2)]^2$$

The result is an expression with two inner sets of parentheses. Focus on the first of these two sets, evaluating the power and then the subtraction:

$$= [(123 - 121)^4 - (6^2 \div 2^2)]^2$$

Work on the remaining inner set of parentheses, evaluating the two powers from left to right and then the division:

$$= [2^4 - (36 \div 2^2)]^2$$

$$= [2^4 - (36 \div 4)]^2$$

$$= [2^4 - 9]^2$$

Now evaluate what's left inside the parentheses, evaluating the power and then the subtraction:

$$= [16 - 9]^2$$

$$= 7^2$$

Finish by evaluating the power: $7^2 = 49$.

Chapter 5

Dividing Attention: Divisibility, Factors, and Multiples

This chapter provides an important bridge between the Big Four operations earlier in the book and the topic of fractions coming up. And right up front, here's the big secret: Fractions are really just division. So, before you tackle fractions, the focus here is on *divisibility* — when you can evenly divide one number by another without getting a remainder.

First, I show you a few handy tricks for finding out whether one number is divisible by another — without actually having to divide. Next, I introduce the concepts of *factors* and *multiples,* which are closely related to divisibility.

The next few sections focus on factors. I show you how to distinguish between *prime numbers* (numbers that have exactly two factors) and *composite numbers* (numbers that have three or more factors). Next, I show you how to find all the factors of a number and how to break a number down into its prime factors. Most important, I show how to find the *greatest common factor* (GCF) of a set of numbers.

After that, I discuss multiples in detail. You discover how to generate multiples of a number, and I also give you a method for determining the *least common multiple* (LCM) of a set of numbers. By the end of this chapter, you should be well equipped to divide and conquer the fractions in Chapters 6 and 7.

Checking for Leftovers: Divisibility Tests

When one number is *divisible* by a second number, you can divide the first number by the second without having anything left over. For example, 16 is divisible by 8 because $16 \div 8 = 2$, with no remainder. You can use a bunch of tricks for testing divisibility without actually doing the division.

The most common tests are for divisibility by 2, 3, 5, and 11. Testing for divisibility by 2 and 5 are child's play; testing for divisibility by 3 and 11 requires a little more work. Here are some quick tests:

- ✔ **By 2:** Any number that ends in an even number (2, 4, 6, 8, or 0) is even — that is, it's divisible by 2. All numbers ending in an odd number (1, 3, 5, 7, or 9) are odd — that is, they aren't divisible by 2.

- ✔ **By 3:** Any number whose digital root is 3, 6, or 9 is divisible by 3; all other numbers (except 0) aren't. To find the *digital root* of a number, just add up the digits. If the result has more than one digit, add up *those* digits and repeat until your result has one digit.

- ✔ **By 5:** Any number that ends in 5 or 0 is divisible by 5; all other numbers aren't.

- ✔ **By 11:** Alternately place plus signs and minus signs in front of all the digits and find the answer. If the result is 0 or any number that's divisible by 11 (even if this result is a negative number), the number is divisible by 11; otherwise, it isn't. **Remember:** Always put a plus sign in front of the *first number*.

Q. Which of the following numbers are divisible by 3?

a. 31

b. 54

c. 768

d. 2,809

A. Add the digits to determine each number's digital root; if the digital root is 3, 6, or 9, the number is divisible by 3:

a. 31: **No,** because 3 + 1 = 4 (check: 31 ÷ 3 = 10 r 1)

b. 54: **Yes,** because 5 + 4 = 9 (check: 54 ÷ 3 = 18)

c. 768: **Yes,** because 7 + 6 + 8 = 21 and 2 + 1 = 3 (check: 768 ÷ 3 = 256)

d. 2,809: **No,** because 2 + 8 + 0 + 9 = 19, 1 + 9 = 10, and 1 + 0 = 1 (check: 2,809 ÷ 3 = 936 r 1)

Q. Which of the following numbers are divisible by 11?

a. 71

b. 154

c. 528

d. 28,094

A. Place + and – signs between the numbers and determine whether the result is 0 or a multiple of 11:

a. 71: **No,** because +7 – 1 = 6 (check: 71 ÷ 11 = 6 r 5)

b. 154: **Yes,** because +1 – 5 + 4 = 0 (check: 154 ÷ 11 = 14)

c. 528: **Yes,** because +5 – 2 + 8 = 11 (check: 528 ÷ 11 = 48)

d. 28,094: **Yes,** because +2 – 8 + 0 – 9 + 4 = –11 (check: 28,094 ÷ 11 = 2,554)

1. Which of the following numbers are divisible by 2?

   a. 37

   b. 82

   c. 111

   d. 75,316

Solve It

2. Which of the following numbers are divisible by 5?

   a. 75

   b. 103

   c. 230

   d. 9,995

Solve It

3. Which of the following numbers are divisible by 3?

   a. 81

   b. 304

   c. 986

   d. 4,444,444

Solve It

4. Which of the following numbers are divisible by 11?

   a. 42

   b. 187

   c. 726

   d. 1,969

Solve It

Understanding Factors and Multiples

In the preceding section, I introduce the concept of divisibility. For example, 12 is divisible by 3 because $12 \div 3 = 4$, with no remainder. You can also describe this relationship between 12 and 3 using the words *factor* and *multiple*. When you're working with positive numbers, the factor is always the *smaller number* and the multiple is the *bigger number*. For example, 12 is divisible by 3, so

✔ The number 3 is a *factor* of 12.

✔ The number 12 is a *multiple* of 3.

Q. The number 40 is divisible by 5, so which number is the factor and which is the multiple?

A. The number **5 is the factor and 40 is the multiple,** because 5 is smaller and 40 is larger.

Q. Which two of the following statements means the same thing as "18 is a multiple of 6"?

a. 6 is a factor of 18.

b. 18 is divisible by 6.

c. 6 is divisible by 18.

d. 18 is a factor of 6.

A. **Choices *a* and *b*.** The number 6 is the factor and 18 is the multiple, because 6 is smaller than 18, so *a* is correct. And $18 \div 6 = 3$, so 18 is divisible by 6; therefore, *b* is correct.

---

5. Which of the following statements are true, and which are false?

a. 5 is a factor of 15.

b. 9 is a multiple of 3.

c. 11 is a factor of 12.

d. 7 is a multiple of 14.

Solve It

6. Which two of these statements mean the same thing as "18 is divisible by 6"?

a. 18 is a factor of 6.

b. 18 is a multiple of 6.

c. 6 is a factor of 18.

d. 6 is a multiple of 18.

Solve It

7. Which two of these statements mean the same thing as "10 is a factor of 50"?

   a. 10 is divisible by 50.

   b. 10 is a multiple of 50.

   c. 50 is divisible by 10.

   d. 50 is a multiple of 10.

Solve It

8. Which of the following statements are true, and which are false?

   a. 3 is a factor of 42.

   b. 11 is a multiple of 121.

   c. 88 is a multiple of 9.

   d. 11 is a factor of 121.

Solve It

One Number, Indivisible: Identifying Prime (And Composite) Numbers

Every counting number greater than 1 is either a prime number or a composite number. A *prime number* has exactly two factors — 1 and the number itself. For example, the number 5 is prime because its only two factors are 1 and 5. A *composite number* has at least three factors. For example, the number 4 has three factors: 1, 2, and 4. (**Note:** The number 1 is the only counting number that isn't prime or composite, because its *only* factor is 1.) The first six prime numbers are 2, 3, 5, 7, 11, and 13.

When testing to see whether a number is prime or composite, perform divisibility tests in the following order (from easiest to hardest): 2, 5, 3, 11, 7, and 13. If you find that a number is divisible by one of these, you know that it's composite and you don't have to perform the remaining tests. Here's how you know which tests to perform:

  ✔ If a number less than 121 isn't divisible by 2, 3, 5, or 7, it's prime; otherwise, it's composite.

  ✔ If a number less than 289 isn't divisible by 2, 3, 5, 7, 11, or 13, it's prime; otherwise, it's composite.

Remember that 2 is the only prime number that's even. The next three odd numbers are prime — 3, 5, and 7. To keep the list going, think "lucky 7, lucky 11, unlucky 13" — they're all prime.

Q. For each of the following numbers, tell which is prime and which is composite.

    a. 185

    b. 243

    c. 253

    d. 263

A. Check divisibility to identify prime and composite numbers:

    a. 185 is composite. The number 185 ends in 5, so it's divisible by 5.

    b. 243 is composite. The number 243 ends in an odd number, so it isn't divisible by 2. It doesn't end in 5 or 0, so it isn't divisible by 5. Its digital root is 9 (because 2 + 4 + 3 = 9), so it's divisible by 3. The math shows you that 243 ÷ 3 = 81.

    c. 253 is composite. The number 253 ends in an odd number, so it isn't divisible by 2. It doesn't end in 5 or 0, so it isn't divisible by 5. Its digital root is 1 (because 2 + 5 + 3 = 10 and 1 + 0 = 1), so it isn't divisible by 3. But it is divisible by 11, because it passes the + and − test (+ 2 − 5 + 3 = 0). If you do the math, you find that 253 = 11 × 23.

    d. 263 is prime. The number 263 ends in an odd number, so it isn't divisible by 2. It doesn't end in 5 or 0, so it isn't divisible by 5. Its digital root is 2 (because 2 + 6 + 3 = 11 and 1 + 1 = 2), so it isn't divisible by 3. It isn't divisible by 11, because it fails the + and − test (+2 − 6 + 3 = −1, which isn't 0 or divisible by 11). It isn't divisible by 7, because 263 ÷ 7 = 37 r 2. And it isn't divisible by 13, because 263 ÷ 13 = 20 r 3.

---

9. Which of the following numbers are prime, and which are composite?

    a. 3

    b. 9

    c. 11

    d. 14

Solve It

10. Of the following numbers, tell which are prime and which are composite.

    a. 65

    b. 73

    c. 111

    d. 172

Solve It

11. Find out whether each of these numbers is prime or composite.

   a. 23

   b. 51

   c. 91

   d. 113

Solve It

12. Figure out which of the following are prime numbers and which are composite numbers.

   a. 143

   b. 169

   c. 187

   d. 283

Solve It

Generating a Number's Factors

When one number is divisible by a second number, that second number is a *factor* of the first. For example, 10 is divisible by 2, so 2 is a factor of 10.

A good way to find all the factors of a number is by finding all of that number's factor pairs. A *factor pair* of a number is any pair of two numbers that, when multiplied together, equal that number. For example, 30 has four factor pairs — 1×30, 2×15, 3×10, and 5×6 — because

$$1 \times 30 = 30$$
$$2 \times 15 = 30$$
$$3 \times 10 = 30$$
$$5 \times 6 = 30$$

Here's how to find *all* the factor pairs of a number:

1. **Begin the list with 1 times the number itself.**

2. **Try to find a factor pair that includes 2 — that is, see whether the number is divisible by 2 (for tricks on testing for divisibility, see Chapter 7).**

   If it is, list the factor pair that includes 2.

3. **Test the number 3 in the same way.**

4. **Continue testing numbers until you find no more factor pairs.**

Q. Write down all the factor pairs of 18.

A. **1 × 18, 2 × 9, 3 × 6.** According to Step 1, begin with 1 × 18:

> 1 × 18

The number 18 is even, so it's divisible by 2. And 18 ÷ 2 = 9, so the next factor pair is 2 × 9:

> 1 × 18
>
> 2 × 9

The digital root of 18 is 9 (because 1 + 8 = 9), so 18 is divisible by 3. And 18 ÷ 3 = 6, so the next factor pair is 3 × 6:

> 1 × 18
>
> 2 × 9
>
> 3 × 6

The number 18 isn't divisible by 4, because 18 ÷ 4 = 4 r 2. And 18 isn't divisible by 5, because it doesn't end with 5 or 0. This list of factor pairs is complete because there are no more numbers between 3 and 6, the last factor pair on the list.

---

13. Find all the factor pairs of 12.

Solve It

14. Write down all the factor pairs of 28.

Solve It

---

15. Figure out all the factor pairs of 40.

Solve It

16. Find all the factor pairs of 66.

Solve It

Decomposing a Number into Its Prime Factors

Every number is the product of a unique set of *prime factors,* a group of prime numbers (including repeats) that, when multiplied together, equals that number. This section shows you how to find those prime factors for a given number, a process called *decomposition.*

An easy way to decompose a number is to make a factorization tree. Here's how:

1. **Find two numbers that multiply to equal the original number; write them as numbers that branch off the original one.**

   Knowing the multiplication table can often help you here.

2. **If either number is prime, circle it and end that branch.**

3. **Continue branching off non-prime numbers into two factors; whenever a branch reaches a prime number, circle it and close the branch.**

   When every branch ends in a circled number, you're finished — just gather up the circled numbers.

Q. Decompose the number 48 into its prime factors.

A. **48 = 2 × 2 × 2 × 2 × 3.** Begin making a factorization tree by finding two numbers that multiply to equal 48:

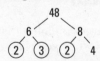

Continue making branches of the tree by doing the same for 6 and 8:

Circle the prime numbers and close those branches. At this point, the only open branch is 4. Break it down into 2 and 2:

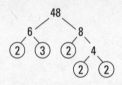

Every branch ends in a circled number, so you're finished. The prime factors are 2, 2, 2, 2, and 3.

17. Decompose 18 into its prime factors.

Solve It

18. Decompose 42 into its prime factors.

Solve It

19. Decompose 81 into its prime factors.

Solve It

20. Decompose 120 into its prime factors.

Solve It

Finding the Greatest Common Factor

The *greatest common factor* (GCF) of a set of numbers is the largest number that's a factor of every number in that set. Finding the GCF is helpful when you want to reduce a fraction to its lowest terms (see Chapter 6).

You can find the GCF in two ways. The first option is to list all the factor pairs of the numbers and choose the largest factor that appears in both (or all) the lists. (For info on finding factor pairs, see the earlier section "Generating a Number's Factors.")

The other method uses prime factors, which I discuss in the preceding section. Here's how to find the GCF:

1. **Decompose the numbers into their prime factors.**

2. **Underline the factors that all the original numbers have in common.**

3. **Multiply the underlined numbers to get the GCF.**

Q. Find the greatest common factor of 12 and 20.

A. **4.** Write down all the factor pairs of 12 and 20:

Factor pairs of 12: 1 × 12, 2 × 6, 3 × 4

Factor pairs of 20: 1 × 20, 2 × 10, 4 × 5

The number 4 is the greatest number that appears in both lists of factor pairs, so it's the GCF.

Q. Find the greatest common factor of 24, 36, and 42.

A. **6.** Decompose all three numbers down to their prime factors:

$24 = 2 \times 2 \times 2 \times 3$

$36 = 2 \times 2 \times 3 \times 3$

$42 = 2 \times 3 \times 7$

Underline all factors that are common to all three numbers:

$24 = \underline{2} \times 2 \times 2 \times \underline{3}$

$36 = \underline{2} \times 2 \times \underline{3} \times 3$

$42 = \underline{2} \times \underline{3} \times 7$

Multiply those underlined numbers to get your answer:

$2 \times 3 = 6$

---

21. Find the greatest common factor of 10 and 22.

Solve It

22. What's the GCF of 8 and 32?

Solve It

---

23. Find the GCF of 30 and 45.

Solve It

24. Figure out the GCF of 27 and 72.

Solve It

25. Find the GCF of 15, 20, and 35.

Solve It

26. Figure out the GCF of 44, 56, and 72.

Solve It

Generating the Multiples of a Number

Generating the multiples of a number is easier than generating the factors: Just multiply the number by 1, 2, 3, and so forth. But unlike the factors of a number — which are always less than the number itself — in the positive numbers, the multiples of a number are greater than or equal to that number. Therefore, you can never write down all the multiples of any positive number.

Q. Find the first six (positive) multiples of 4.

A. **4, 8, 12, 16, 20, 24.** Write down the number 4 and keep adding 4 to it until you've written down six numbers.

Q. List the first six multiples of 12.

A. **12, 24, 36, 48, 60, 72.** Write down 12 and keep adding 12 to it until you've written down six numbers.

27. Write the first six multiples of 5.

Solve It

28. Generate the first six multiples of 7.

Solve It

29. List the first ten multiples of 8.

Solve It

30. Write the first six multiples of 15.

Solve It

Finding the Least Common Multiple

The *least common multiple* (LCM) of a set of numbers is the smallest number that's a multiple of every number in that set. For small numbers, you can simply list the first several multiples of each number until you get a match.

When you're finding the LCM of two numbers, you may want to list the multiples of the larger number first, stopping when the number of multiples you've written down equals the smaller number. Then list the multiples of the smaller number and look for a match.

However, you may have to write down a lot of multiples with this method, and the disadvantage becomes even greater when you're trying to find the LCM of large numbers. I recommend a method that uses prime factors when you're facing big numbers or more than two numbers. Here's how:

1. **Write down the prime decompositions of all the numbers.**

   See the earlier "Decomposing a Number into Its Prime Factors" section for details.

2. **For each prime number you find, underline the *most repeated occurrences* of each.**

   In other words, compare the decompositions. If one breakdown contains two 2s and another contains three 2s, you'd underline the three 2s. If one decomposition contains one 7 and the rest don't have any, you'd underline the 7.

3. **Multiply the underlined numbers to get the LCM.**

Q. Find the LCM of 6 and 8.

A. **24.** Because 8 is the larger number, write down six multiples of 8:

Multiples of 8: 8, 16, 24, 32, 40, 48

Now, write down multiples of 6 until you find a matching number:

Multiples of 6: 6, 12, 18, 24

Q. Find the LCM of 12, 15, and 18.

A. **180.** Begin by writing the prime decompositions of all three numbers. Then, for each prime number you find, underline the most repeated occurrences of each:

$12 = \underline{2} \times \underline{2} \times 3$

$15 = 3 \times \underline{5}$

$18 = 2 \times \underline{3} \times \underline{3}$

Notice that 2 appears in the decomposition of 12 most often (twice), so I underline both of those 2s. Similarly, 3 appears in the decomposition of 18 most often (twice), and 5 appears in the decomposition of 15 most often (once). Now, multiply all the underlined numbers:

$2 \times 2 \times 3 \times 3 \times 5 = 180$

31. Find the LCM of 4 and 10.

Solve It

32. Find the LCM of 7 and 11.

Solve It

33. Find the LCM of 9 and 12.

Solve It

34. Find the LCM of 18 and 22.

Solve It

Solutions to Divisibility, Factors, and Multiples

The following are the answers to the practice questions presented earlier in this chapter.

1 Which of the following numbers are divisible by 2?

a. 37: **No,** because it's odd (check: 37 ÷ 2 = 18 r 1)

b. 82: **Yes,** because it's even (check: 82 ÷ 2 = 41)

c. 111: **No,** because it's odd (check: 111 ÷ 2 = 55 r 1)

d. 75,316: **Yes,** because it's even (check: 75,316 ÷ 2 = 37,658)

2 Which of the following numbers are divisible by 5?

a. 75: **Yes,** because it ends in 5 (check: 75 ÷ 5 = 25)

b. 103: **No,** because it ends in 3, not 0 or 5 (check: 103 ÷ 5 = 20 r 3)

c. 230: **Yes,** because it ends in 0 (check: 230 ÷ 5 = 46)

d. 9,995: **Yes,** because it ends in 5 (check: 9,995 ÷ 5 = 1,999)

3 Which of the following numbers are divisible by 3? ***Note:*** The digital root ends in 3, 6, or 9 for numbers divisible by 3:

a. 81: **Yes,** because 8 + 1 = 9 (check: 81 ÷ 3 = 27)

b. 304: **No,** because 3 + 0 + 4 = 7 (check: 304 ÷ 3 = 101 r 1)

c. 986: **No,** because 9 + 8 + 6 = 23 and 2 + 3 = 5 (check: 986 ÷ 3 = 328 r 2)

d. 4,444,444: **No,** because 4 + 4 + 4 + 4 + 4 + 4 + 4 = 28, 2 + 8 = 10, and 1 + 0 = 1 (check: 4,444,444 ÷ 3 = 1,481,481 r 1)

4 Which of the following numbers are divisible by 11? ***Note:*** Answers add up to 0 or a multiple of 11 for numbers divisible by 11:

a. 42: **No,** because +4 – 2 = 2 (check: 42 ÷ 11 = 3 r 9)

b. 187: **Yes,** because +1 – 8 + 7 = 0 (check: 187 ÷ 11 = 17)

c. 726: **Yes,** because +7 – 2 + 6 = 11 (check: 726 ÷ 11 = 66)

d. 1,969: **Yes,** because +1 – 9 + 6 – 9 = –11 (check: 1,969 ÷ 11 = 179)

5 Which of the following statements are true, and which are false?

a. 5 is a factor of 15. **True:** 5 × 3 = 15.

b. 9 is a multiple of 3. **True:** 3 × 3 = 9.

c. 11 is a factor of 12. **False:** You can't multiply 11 by any whole number to get 12.

d. 7 is a multiple of 14. **False:** The number 7 is a factor of 14.

6 Which of these statements means the same thing as "18 is divisible by 6"? **Choices *b* and *c*.** You're looking for something that says that the smaller number, 6, is a factor of the larger number (choice *c*) or one that says the larger number, 18, is a multiple of the smaller number (choice *b*).

7 Which of these statements means the same thing as "10 is a factor of 50"? **Choices *c* and *d*.** Factors are numbers you multiply to get larger ones, so you can say 50 is divisible by 10 (choice *c*). Multiples are the larger numbers, the products you get when you multiply two factors; you can say 50 is a multiple of 10 (choice *d*) because $10 \times 5 = 50$.

8 Which of the following statements are true, and which are false?

a. 3 is a factor of 42. **True:** $3 \times 14 = 42$.

b. 11 is a multiple of 121. **False:** The number 11 is a factor of 121.

c. 88 is a multiple of 9. **False:** You can't multiply 9 by any whole numbers to get 88 because $88 \div 9 = 9 \text{ r } 7$.

d. 11 is a factor of 121. **True:** $11 \times 11 = 121$.

9 Which of the following numbers are prime, and which are composite?

a. 3 is prime. The only factors of 3 are 1 and 3.

b. 9 is composite. The factors of 9 are 1, 3, and 9.

c. 11 is prime. Eleven's only factors are 1 and 11.

d. 14 is composite. As an even number, 14 is also divisible by 2 and therefore can't be prime.

10 Of the following numbers, tell which are prime and which are composite.

a. 65 is composite. Because 65 ends in 5, it's divisible by 5.

b. 73 is prime. The number 73 isn't even, doesn't end in 5 or 0, and isn't a multiple of 7.

c. 111 is composite. The digital root of 111 is $1 + 1 + 1 = 3$, so it's divisible by 3 (check: $111 \div 3 = 37$).

d. 172 is composite. The number 172 is even, so it's divisible by 2.

11 Find out whether each of these numbers is prime or composite.

a. 23 is prime. The number 23 isn't even, doesn't end in 5 or 0, has a digital root of 5, and isn't a multiple of 7.

b. 51 is composite. The digital root of 51 is 6, so it's a multiple of 3 (check: $51 \div 3 = 17$).

c. 91 is composite. The number 91 is a multiple of 7: $7 \times 13 = 91$.

d. 113 is prime. The number 113 is odd, doesn't end in 5 or 0, and has a digital root of 5, so it's not divisible by 2, 5, or 3. It's also not a multiple of 7: $113 \div 7 = 16 \text{ r } 1$.

12 Figure out which of the following are prime numbers and which are composite numbers:

a. 143 is composite. $+1 - 4 + 3 = 0$, so 143 is divisible by 11.

b. 169 is composite. You can evenly divide 13 into 169 to get 13.

c. 187 is composite. $+1 - 8 + 7 = 0$, so 187 is a multiple of 11.

d. 283 is prime. The number 283 is odd, doesn't end in 5 or 0, and has a digital root of 4; therefore, it's not divisible by 2, 5, or 3. It's not divisible by 11, because $+2 - 8 + 3 = 3$, which isn't a multiple of 11. It also isn't divisible by 7 (because $283 \div 7 = 40 \text{ r } 3$) or 13 (because $283 \div 13 = 21 \text{ r } 10$).

13 The factor pairs of 12 are $\mathbf{1 \times 12}$, $\mathbf{2 \times 6}$, **and** $\mathbf{3 \times 4}$. The first factor pair is 1×12. And 12 is divisible by 2 ($12 \div 2 = 6$), so the next factor pair is 2×6. Because 12 is divisible by 3 ($12 \div 3 = 4$), the next factor pair is 3×4.

14 The factor pairs of 28 are **1 × 28, 2 × 14, and 4 × 7.** The first factor pair is 1 × 28. And 28 is divisible by 2 (28 ÷ 2 = 14), so the next factor pair is 2 × 14. Although 28 isn't divisible by 3, it's divisible by 4 (28 ÷ 4 = 7), so the next factor pair is 4 × 7. Finally, 28 isn't divisible by 5 or 6.

15 The factor pairs of 40 are **1 × 40, 2 × 20, 4 × 8, and 5 × 8.** The first factor pair is 1 × 40. Because 40 is divisible by 2 (40 ÷ 2 = 20), the next factor pair is 2 × 20. Although 40 isn't divisible by 3, it's divisible by 4 (40 ÷ 4 = 10), so the next factor pair is 4 × 10. And 40 is divisible by 5 (40 ÷ 5 = 8), so the next factor pair is 5 × 8. Finally, 40 isn't divisible by 6 or 7.

16 The factor pairs of 66 are **1 × 66, 2 × 33, 3 × 22, and 6 × 11.** The first factor pair is 1 × 66. The number 66 is divisible by 2 (66 ÷ 2 = 33), so the next factor pair is 2 × 33. It's also divisible by 3 (66 ÷ 3 = 22), so the next factor pair is 3 × 22. Although 66 isn't divisible by 4 or 5, it's divisible by 6 (66 ÷ 6 = 11), so the next factor pair is 6 × 11. Finally, 66 isn't divisible by 7, 8, 9, or 10.

17 Decompose 18 into its prime factors. **18 = 2 × 2 × 3.** Here's one possible factoring tree:

18 Decompose 42 into its prime factors. **42 = 2 × 3 × 7.** Here's one possible factoring tree:

19 Decompose 81 into its prime factors. **81 = 3 × 3 × 3 × 3.** Here's one possible factoring tree:

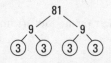

20 Decompose 120 into its prime factors. **120 = 2 × 2 × 2 × 3 × 5.** Here's one possible factoring tree:

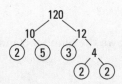

21 The GCF of 10 and 22 is **2.** Write down all the factor pairs of 10 and 22:

      10: 1 × 10, 2 × 5

      22: 1 × 22, 2 × 11

The number 2 is the greatest number that appears on both lists.

22 The GCF of 8 and 32 is **8.** Write down all the factor pairs of 8 and 32:

   8: 1×8, 2×4

   32: 1×32, 2×16, 4×8

The greatest number that appears on both lists is 8.

23 The GCF of 30 and 45 is **15.** Write down all the factor pairs of 30 and 45:

   30: 1×30, 2×15, 3×10, 5×6

   45: 1×45, 3×15, 5×9

The greatest number that appears on both lists is 15.

24 The GCF of 27 and 72 is **9.** Decompose 27 and 72 into their prime factors and underline every factor that's common to both:

   $27 = \underline{3} \times \underline{3} \times 3$

   $72 = 2 \times 2 \times 2 \times \underline{3} \times \underline{3}$

Multiply those underlined numbers to get your answer: $3 \times 3 = 9$.

25 The GCF of 15, 20, and 35 is **5.** Decompose the three numbers into their prime factors and underline every factor that's common to all three:

   $15 = 3 \times \underline{5}$

   $20 = 2 \times 2 \times \underline{5}$

   $35 = \underline{5} \times 7$

The only factor common to all three numbers is 5.

26 The GCF of 44, 56, and 72 is **4.** Decompose all three numbers to their prime factors and underline each factor that's common to all three:

   $44 = \underline{2} \times \underline{2} \times 11$

   $56 = \underline{2} \times \underline{2} \times 2 \times 7$

   $72 = \underline{2} \times \underline{2} \times 2 \times 3 \times 3$

Multiply those underlined numbers to get your answer: $2 \times 2 = 4$.

27 The first six multiples of 5 are **5, 10, 15, 20, 25, and 30.** Write down the number 5 and keep adding 5 to it until you've written six numbers.

28 The first six multiples of 7 are **7, 14, 21, 28, 35, and 42.**

29 The first ten multiples of 8 are **8, 16, 24, 32, 40, 48, 56, 64, 72, and 80.**

30 The first six multiples of 15 are **15, 30, 45, 60, 75, and 90.**

31 The LCM of 4 and 10 is **20.** Write down four multiples of 10:

Multiples of 10: 10, 20, 30, 40

Next, generate multiples of 4 until you find a matching number:

Multiples of 4: 4, 8, 12, 16, 20

32 The LCM of 7 and 11 is **77.** Write down seven multiples of 11:

Multiples of 11: 11, 22, 33, 44, 55, 66, 77

Next, generate multiples of 7 until you find a matching number:

Multiples of 7: 7, 14, 21, 28, 35, 42, 49, 56, 63, 70, 77

33 The LCM of 9 and 12 is **36.** Write down nine multiples of 12:

Multiples of 12: 12, 24, 36, 48, 60, 72, 84, 96, 108

Next, generate multiples of 9 until you find a matching number:

Multiples of 9: 9, 18, 27, 36

34 The LCM of 18 and 22 is **198.** First, decompose both numbers into their prime factors. Then underline the most frequent occurrences of each prime number:

$18 = \underline{2} \times \underline{3} \times \underline{3}$

$22 = 2 \times \underline{11}$

The factor 2 appears only once in any decomposition, so I underline a 2. The number 3 appears twice in the decomposition of 18, so I underline both of these. The number 11 appears only once, in the decomposition of 22, so I underline it. Now, multiply all the underlined numbers:

$2 \times 3 \times 3 \times 11 = 198$

Part II

Slicing Things Up: Fractions, Decimals, and Percents

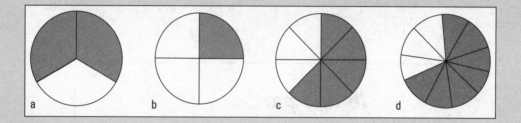

In this part. . .

- ✔ Understand fractions, including improper fractions and mixed numbers.

- ✔ Apply the Big Four operations (addition, subtraction, multiplication, and division) to fractions, decimals, and percents.

- ✔ Change rational numbers to fractions, decimals, or percents.

- ✔ Use fractions, decimals, and percents to solve word problems.

Chapter 6

Fractions Are a Piece of Cake

In This Chapter

- Knowing the basics of fractions
- Converting between improper fractions and mixed numbers
- Increasing and reducing the terms of fractions
- Comparing fractions using cross-multiplication

Fractions come into play when you divide a whole object into equal pieces. For example, if you cut a cake into four equal pieces, each piece is $\frac{1}{4}$ of the whole cake. Fractions are commonly used in everything from cooking to carpentry. You see a lot of them in math classes, too!

This chapter starts off with the basics of fractions, showing you how fractions use two numbers — the *numerator* (top number) and *denominator* (bottom number) — to represent part of a whole object. You discover how to recognize the three basic types of fractions (proper fractions, improper fractions, and mixed numbers) and how to convert back and forth between improper fractions and mixed numbers. Then I get you started increasing and reducing the terms of fractions. Finally, I show you how to cross-multiply a pair of fractions to change them to two new fractions with a common denominator. You use this trick to find out which fraction is greater and which is lesser.

Getting Down the Basic Fraction Stuff

Fractions represent parts of a whole — that is, quantities that fall between the whole numbers. Probably the most commonly used fraction is $\frac{1}{2}$, which is *one-half.* When you cut a cake into two pieces and take one for yourself, you get $\frac{1}{2}$ of the cake — I hope you're hungry! In Figure 6-1, your piece is shaded.

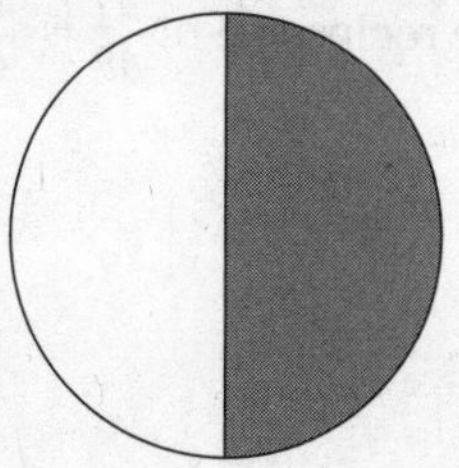

Figure 6-1: Half of a cake.

When you slice yourself a fraction of a cake, that fraction contains two numbers, and each number tells you something different:

✔ The top number — called the *numerator* — tells you the number of *shaded* slices.

✔ The bottom number — called the *denominator* — tells you the *total* number of slices.

When the numerator of a fraction is less than the denominator, that fraction is a *proper fraction*. If the numerator is greater than the denominator, that fraction is an *improper fraction*. You can convert improper fractions into mixed numbers, as I show you in the next section.

Some fractions can be easily written as whole numbers:

✔ When a fraction's denominator is 1, that fraction is equal to its numerator.

✔ When a fraction's numerator and denominator are the same, that fraction is equal to 1. (This idea is important when you want to change the terms of a fraction — see "Increasing and Reducing the Terms of Fractions" for details.)

When you reverse the order of the numerator and denominator in a fraction, the result is the *reciprocal* of that fraction. You use reciprocals to divide by fractions; check out Chapter 7 for more info.

Q. For each cake pictured below, identify the fraction of the cake that's shaded.

a.   b.   c.   d.

A. Put the number of shaded slices over the number of total slices in each cake:

a. $\frac{2}{3}$

b. $\frac{1}{4}$

c. $\frac{5}{8}$

d. $\frac{7}{10}$

Q. What's the reciprocal of each of the following fractions?

a. $\frac{3}{4}$

b. $\frac{6}{11}$

c. $\frac{22}{7}$

d. $\frac{41}{48}$

A. To find the reciprocal, switch around the numerator and the denominator:

a. The reciprocal of $\frac{3}{4}$ is $\frac{4}{3}$.

b. The reciprocal of $\frac{6}{11}$ is $\frac{11}{6}$.

c. The reciprocal of $\frac{22}{7}$ is $\frac{7}{22}$.

d. The reciprocal of $\frac{41}{48}$ is $\frac{48}{41}$.

1. For each cake pictured, identify the
fraction of the cake that's shaded.

a.   b.   c.   d.

Solve It

2. Which of the following fractions are
proper? Which are improper?

a. $\frac{3}{2}$

b. $\frac{8}{9}$

c. $\frac{20}{23}$

d. $\frac{75}{51}$

Solve It

3. Rewrite each of the following fractions as a
whole number:

a. $\frac{3}{3}$

b. $\frac{10}{1}$

c. $\frac{10}{10}$

d. $\frac{81}{1}$

Solve It

4. Find the reciprocal of the following
fractions:

a. $\frac{5}{7}$

b. $\frac{10}{3}$

c. $\frac{12}{17}$

d. $\frac{80}{91}$

Solve It

In Mixed Company: Converting between Mixed Numbers and Improper Fractions

When the numerator (top number) is greater than the denominator (bottom number), that fraction is an *improper fraction*. An alternative form for an improper fraction is as a *mixed number*, which is made up of a whole number and a fraction.

For example, you can represent the improper fraction $\frac{3}{2}$ as the equivalent mixed number $1\frac{1}{2}$. The mixed number $1\frac{1}{2}$ means $1+\frac{1}{2}$. To see why $\frac{3}{2}=1\frac{1}{2}$, realize that *three halves* of a cake is the same as *one* whole cake plus another *half*. Every improper fraction has an equivalent mixed number, and vice versa.

Sometimes at the beginning of a fraction problem, converting a mixed number to an improper fraction makes the problem easier to solve. Here's how to make the switch from mixed number to improper fraction:

1. **Multiply the whole number by the fraction's denominator (bottom number).**

2. **Add the numerator (top number) to the product from Step 1.**

3. **Place the sum from Step 2 over the original denominator.**

Similarly, at the end of some problems, you may need to convert an improper fraction to a mixed number. To do so, simply divide the numerator by the denominator. Then build a mixed number:

✔ The quotient is the whole number.

✔ The remainder is the numerator of the fraction.

✔ The denominator of the fraction stays the same.

Think of the fraction bar as a division sign.

Q. Convert the mixed number $2\frac{3}{4}$ to an improper fraction.

A. $\frac{11}{4}$. Multiply the whole number (2) by the denominator (4), and then add the numerator (3):

$2 \times 4 + 3 = 11$

Use this number as the numerator of your answer, keeping the same denominator:

$\frac{11}{4}$

Q. Convert the mixed number $3\frac{5}{7}$ to an improper fraction:

A. $\frac{26}{7}$. Multiply the whole number (3) by the denominator (7), and then add the numerator (5). This time, I do the whole process in one step:

$$3\frac{5}{7} = \frac{(3\times 7 + 5)}{7} = \frac{26}{7}$$

Q. Convert the improper fraction $\frac{11}{2}$ to a mixed number:

A. $5\frac{1}{2}$. Divide the numerator (11) by the denominator (2):

Quotient →
$$2\overline{)\,11}$$
with 5 above
$$-10$$
Remainder → 1

Now build a mixed number using the quotient (5) as the whole number and the remainder (1) as the numerator, keeping the same denominator (2):

$$5\frac{1}{2}$$

Q. Convert the improper fraction $\frac{39}{5}$ to a mixed number:

A. $7\frac{4}{5}$. Divide the numerator (39) by the denominator (5):

Quotient →
$$5\overline{)\,39}$$
with 7 above
$$-35$$
Remainder → 4

Build your answer using the quotient (7) as the whole number and the remainder (4) as the numerator, keeping the same denominator (5):

$$7\frac{4}{5}$$

5. Convert the mixed number $5\frac{1}{4}$ to an improper fraction.

Solve It

6. Change $7\frac{2}{9}$ to an improper fraction.

Solve It

7. Express the mixed number $10\frac{5}{12}$ as an improper fraction.

Solve It

8. Convert the improper fraction $\frac{13}{4}$ to a mixed number.

Solve It

9. Express the improper fraction $\frac{29}{10}$ as a mixed number.

Solve It

10. Change $\frac{100}{7}$ to a mixed number.

Solve It

Increasing and Reducing the Terms of Fractions

When you cut a cake into two pieces and take one piece, you have $\frac{1}{2}$ of the cake. And when you cut it into four pieces and take two, you have $\frac{2}{4}$ of the cake. Finally, when you cut it into six pieces and take three, you have $\frac{3}{6}$ of the cake. Notice that in all these cases, you get the same amount of cake. This shows you that the fractions $\frac{1}{2}$, $\frac{2}{4}$, and $\frac{3}{6}$ are *equal;* so are the fractions $\frac{10}{20}$ and $\frac{1,000,000}{2,000,000}$.

Most of the time, writing this fraction as $\frac{1}{2}$ is preferred because the numerator and denominator are the smallest possible numbers. In other words, the fraction $\frac{1}{2}$ is written in lowest terms. At the end of a problem, you often need to reduce a fraction, or write it in lowest terms. There are two ways to do this — the informal way and the formal way:

✔ The informal way to reduce a fraction is to divide both the numerator and the denominator by the same number.

Advantage: The informal way is easy.

Disadvantage: It doesn't always reduce the fraction to lowest terms (though you do get the fraction in lowest terms if you divide by the greatest common factor, which I discuss in Chapter 5).

✔ The formal way is to decompose both the numerator and the denominator into their prime factors and then cancel common factors.

Advantage: The formal way always reduces the fraction to lowest terms.

Disadvantage: It takes longer than the informal way.

TIP

Start every problem using the informal way. If the going gets rough and you're still not sure whether your answer is reduced to lowest terms, switch over to the formal way.

Sometimes at the beginning of a fraction problem, you need to *increase* the terms of a fraction — that is, write that fraction using a greater numerator and denominator. To increase terms, multiply both the numerator and denominator by the same number.

Q. Increase the terms of the fraction $\frac{4}{5}$ to a new fraction whose denominator is 15:

A. $\frac{12}{15}$. To start out, write the problem as follows:

$$\frac{4}{5} = \frac{?}{15}$$

The question mark stands for the numerator of the new fraction, which you want to fill in. Now divide the larger denominator (15) by the smaller denominator (5).

$$15 \div 5 = 3$$

Multiply this result by the numerator:

$$3 \times 4 = 12$$

Finally, take this number and use it to replace the question mark:

$$\frac{4}{5} = \frac{12}{15}$$

Q. Reduce the fraction $\frac{18}{42}$ to lowest terms.

A. $\frac{3}{7}$. The numerator and denominator aren't too large, so use the informal way: To start out, try to find a small number that the numerator and denominator are both divisible by. In this case, notice that the numerator and denominator are both divisible by 2, so divide both by 2:

$$\frac{18}{42} = \frac{(18 \div 2)}{(42 \div 2)} = \frac{9}{21}$$

Next, notice that the numerator and denominator are both divisible by 3 (see Chapter 5 for more on how to tell whether a number is divisible by 3), so divide both by 3:

$$\frac{9}{21} = \frac{(9 \div 3)}{(21 \div 3)} = \frac{3}{7}$$

At this point, there's no number (except for 1) that evenly divides both the numerator and denominator, so this is your answer.

Q. Reduce the fraction $\frac{135}{196}$ to lowest terms.

A. $\frac{135}{196}$. The numerator and denominator are both over 100, so use the formal way. First, decompose both the numerator and denominator down to their prime factors:

$$\frac{135}{196} = \frac{(3 \times 3 \times 3 \times 5)}{(2 \times 2 \times 7 \times 7)}$$

The numerator and denominator have no common factors, so the fraction is already in lowest terms.

11. Increase the terms of the fraction $\frac{2}{3}$ so that the denominator is 18.

Solve It

12. Increase the terms of $\frac{4}{9}$, changing the denominator to 54.

Solve It

13. Reduce the fraction $\frac{12}{60}$ to lowest terms.

Solve It

14. Reduce $\frac{45}{75}$ to lowest terms.

Solve It

15. Reduce the fraction $\frac{135}{180}$ to lowest terms.

Solve It

16. Reduce $\frac{108}{217}$ to lowest terms.

Solve It

Comparing Fractions with Cross-Multiplication

Cross-multiplication is a handy tool for getting a common denominator for two fractions, which is important for many operations involving fractions. In this section, I show you how to cross-multiply to compare a pair of fractions to find out which is greater or less. (In Chapter 7, I show you how to use cross-multiplication to add fractions, and in Chapter 15, I show you how to use it to help solve algebra equations.)

Here's how to cross-multiply two fractions:

1. **Multiply the numerator (top number) of the first fraction by the denominator (bottom number) of the second, writing the answer below the first fraction.**

2. **Multiply the numerator of the second fraction by the denominator of the first, writing the answer below the second fraction.**

The result is that each fraction now has a new number written underneath it. The larger number is below the larger fraction.

You can use cross-multiplication to rewrite a pair of fractions as two new fractions with a common denominator:

1. **Cross-multiply the two fractions to find the numerators of the new fractions.**

2. **Multiply the denominators of the two fractions to find the new denominators.**

When two fractions have the same denominator, the one with the greater numerator is the greater fraction.

Q. Which fraction is greater: $\frac{5}{8}$ or $\frac{6}{11}$?

A. $\frac{5}{8}$. Cross-multiply the two fractions:

$$\frac{5}{8} \diagup\!\!\!\!\diagdown \frac{6}{11}$$
$$55 \qquad 48$$

Because 55 is greater than 48, $\frac{5}{8}$ is greater than $\frac{6}{11}$.

Q. Which of these three fractions is the least: $\frac{3}{4}$, $\frac{7}{10}$, or $\frac{8}{11}$?

A. $\frac{7}{10}$. Cross-multiply the first two fractions:

$$\frac{3}{4} \diagup\!\!\!\!\diagdown \frac{7}{10}$$
$$30 \qquad 28$$

Because 28 is less than 30, $\frac{7}{10}$ is less than $\frac{3}{4}$, so you can rule out $\frac{3}{4}$. Now compare $\frac{7}{10}$ and $\frac{8}{11}$ similarly:

$$\frac{7}{10} \diagup\!\!\!\!\diagdown \frac{8}{11}$$
$$77 \qquad 80$$

Because 77 is less than 80, $\frac{7}{10}$ is less than $\frac{8}{11}$. Therefore, $\frac{7}{10}$ is the least of the three fractions.

17. Which is the greater fraction: $\frac{1}{5}$ or $\frac{2}{9}$?

Solve It

18. Find the lesser fraction: $\frac{3}{7}$ or $\frac{5}{12}$.

Solve It

19. Among these three fractions, which is greatest: $\frac{1}{10}$, $\frac{2}{21}$, or $\frac{3}{29}$?

Solve It

20. Figure out which of the following fractions is the least: $\frac{1}{3}$, $\frac{2}{7}$, $\frac{4}{13}$, or $\frac{8}{25}$.

Solve It

Working with Ratios and Proportions

A *ratio* is a mathematical comparison of two numbers, based on division. For example, suppose you bring 3 shirts and 5 ties with you on a business trip. Here are a few ways to express the ratio of shirts to ties:

3:5   3 to 5 $\dfrac{3}{5}$

A good way to work with a ratio is to turn it into a fraction. Be sure to keep the order the same: The first number goes on top of the fraction, and the second number goes on the bottom.

You can use a ratio to solve problems by setting up a *proportion equation* — that is, an equation involving two ratios.

Q. Clarence has 1 daughter and 4 sons. Set up a proportion equation based on this ratio.

A. $\dfrac{\text{Daughters}}{\text{Sons}} = \dfrac{1}{4}$.

Q. An English language school has 3:7 ratio of European to Asian students. If the school has 28 students from Asia, how many students in the school are from Europe?

A. **12 students.** To begin, set up a proportion based on the ratio of European to Asian students (be sure that the order of the two numbers is the same as the two attributes that they stand for — 3 Europeans and 7 Asians):

$$\frac{\text{Europe}}{\text{Asia}} = \frac{3}{7}$$

Now, increase the terms of the fraction $\dfrac{3}{7}$ so that the number representing the count of Asian students becomes 28:

$$\frac{\text{Europe}}{\text{Asia}} = \frac{3 \times 4}{7 \times 4} = \frac{12}{28}$$

Therefore, given that the school has 28 Asian students, it has 12 European students.

21. A farmers market sells a 4 to 5 ratio of vegetables to fruit. If it sells 35 different types of fruits, how many different types of vegetables does it sell?

Solve It

22. An art gallery is currently featuring an exhibition with a 2:7 ratio of sculpture to paintings. If the exhibition includes 18 sculptures, how many paintings does it include?

Solve It

23. A summer camp has a 7 to 9 ratio of girls to boys. If it has 117 boys, what is the total number of children attending the summer camp?

Solve It

24. The budget of a small town has a 3:8 ratio of state funding to municipal funding. If the town received $600,000 in municipal funding last year, what was its total budget from both state and municipal sources?

Solve It

Solutions to Fractions Are a Piece of Cake

The following are the answers to the practice questions presented in this chapter.

1 For each cake pictured, identify the fraction of the cake that's shaded.

  a. You have 1 shaded slice and 3 slices in total, so it's $\frac{1}{3}$.

  b. You have 3 shaded slices and 4 slices in total, so it's $\frac{3}{4}$.

  c. You have 5 shaded slices and 6 slices in total, so it's $\frac{5}{6}$.

  d. You have 7 shaded slices and 12 slices in total, so it's $\frac{7}{12}$.

2 Which of the following fractions are proper? Which are improper?

  a. The numerator (3) is greater than the denominator (2), so $\frac{3}{2}$ is an **improper fraction.**

  b. The numerator (8) is less than the denominator (9), so $\frac{8}{9}$ is a **proper fraction.**

  c. The numerator (20) is less than the denominator (23), so $\frac{20}{23}$ is a **proper fraction.**

  d. The numerator (75) is greater than the denominator (51), so $\frac{75}{51}$ is an **improper fraction.**

3 Rewrite each of the following fractions as a whole number.

  a. The numerator and denominator are the same, so $\frac{3}{3} = \mathbf{1.}$

  b. The denominator is 1, so $\frac{10}{1} = \mathbf{10.}$

  c. The numerator and denominator are the same, so $\frac{10}{10} = \mathbf{1.}$

  d. The denominator is 1, so $\frac{81}{1} = \mathbf{81.}$

4 Find the reciprocal of the following fractions by switching the numerator and denominator.

  a. The reciprocal of $\frac{5}{7}$ is $\mathbf{\frac{7}{5}}$.

  b. The reciprocal of $\frac{10}{3}$ is $\mathbf{\frac{3}{10}}$.

  c. The reciprocal of $\frac{12}{17}$ is $\mathbf{\frac{17}{12}}$.

  d. The reciprocal of $\frac{80}{91}$ is $\mathbf{\frac{91}{80}}$.

5 $5\frac{1}{4} = \frac{(5 \times 4 + 1)}{4} = \mathbf{\frac{21}{4}}$

6 $7\frac{2}{9} = \frac{(7 \times 9 + 2)}{9} = \mathbf{\frac{65}{9}}$

7 $10\frac{5}{12} = \frac{(10 \times 12 + 5)}{12} = \mathbf{\frac{125}{12}}$

8 $\frac{13}{4} = 3\frac{1}{4}$. Divide the numerator (13) by the denominator (4):

$$
\begin{array}{r}
\text{Quotient} \quad \rightarrow \quad 3 \\
4\overline{)\,13} \\
-12 \\
\hline
\text{Remainder} \quad \rightarrow \quad 1
\end{array}
$$

Build your answer using the quotient (3) as the whole number and the remainder (1) as the numerator, keeping the same denominator (4): $3\frac{1}{4}$.

9 $\frac{29}{10} = 2\frac{9}{10}$. Divide the numerator (29) by the denominator (10):

$$
\begin{array}{r}
\text{Quotient} \quad \rightarrow \quad 2 \\
10\overline{)\,29} \\
-20 \\
\hline
\text{Remainder} \quad \rightarrow \quad 9
\end{array}
$$

Build your answer using the quotient (2) as the whole number and the remainder (9) as the numerator, keeping the same denominator (10): $2\frac{9}{10}$.

10 $\frac{100}{7} = 14\frac{2}{7}$. Divide the numerator (100) by the denominator (7):

$$
\begin{array}{r}
\text{Quotient} \quad \rightarrow \quad 14 \\
7\overline{)\,100} \\
-7 \\
\hline
30 \\
-28 \\
\hline
\text{Remainder} \quad \rightarrow \quad 2
\end{array}
$$

Build your answer using the quotient (14) as the whole number and the remainder (2) as the numerator, keeping the same denominator (7): $14\frac{2}{7}$

11 $\frac{2}{3} = \frac{12}{18}$. To start out, write the problem as follows:

$$\frac{2}{3} = \frac{?}{18}$$

Divide the larger denominator (18) by the smaller denominator (3) and then multiply this result by the numerator (2):

$6 \times 2 = 12$

Take this number and use it to replace the question mark; your answer is $\frac{12}{18}$.

12 $\frac{4}{9} = \frac{\mathbf{24}}{\mathbf{54}}$. Write the problem as follows:

$$\frac{4}{9} = \frac{?}{54}$$

Divide the larger denominator (54) by the smaller denominator (9) and then multiply this result by the numerator (4):

$6 \times 4 = 24$

Take this number and use it to replace the question mark; your answer is $\frac{24}{54}$.

13 $\frac{12}{60} = \frac{\mathbf{1}}{\mathbf{5}}$. The numerator (12) and the denominator (60) are both even, so divide both by 2:

$$\frac{12}{60} = \frac{6}{30}$$

They're still both even, so divide both by 2 again:

$$= \frac{3}{15}$$

Now the numerator and denominator are both divisible by 3, so divide both by 3:

$$= \frac{1}{5}$$

14 $\frac{45}{75} = \frac{\mathbf{3}}{\mathbf{5}}$. The numerator (45) and the denominator (75) are both divisible by 5, so divide both by 5:

$$\frac{45}{75} = \frac{9}{15}$$

Now the numerator and denominator are both divisible by 3, so divide both by 3:

$$= \frac{3}{5}$$

15 $\frac{135}{180} = \frac{\mathbf{3}}{\mathbf{4}}$. The numerator (135) and the denominator (180) are both divisible by 5, so divide both by 5:

$$\frac{135}{180} = \frac{27}{36}$$

Now the numerator and denominator are both divisible by 3, so divide both by 3:

$$= \frac{9}{12}$$

They're still both divisible by 3, so divide both by 3 again:

$$= \frac{3}{4}$$

16 $\frac{108}{217} = \frac{\mathbf{108}}{\mathbf{217}}$. With a numerator and denominator this large, reduce using the formal way. First, decompose both the numerator and denominator down to their prime factors:

$$\frac{108}{217} = \frac{(2 \times 2 \times 3 \times 3 \times 3)}{(7 \times 31)}$$

The numerator and denominator have no common factors, so the fraction is already in lowest terms.

17 $\frac{2}{9}$ **is greater than** $\frac{1}{5}$. Cross-multiply to compare the two fractions:

$$\frac{2}{9} \diagdown\!\!\!\!\diagup \frac{1}{5}$$

10   9

Because 10 is greater than 9, $\frac{2}{9}$ is greater than $\frac{1}{5}$.

18 $\frac{5}{12}$ **is less than** $\frac{3}{7}$. Cross-multiply to compare the two fractions:

$$\frac{5}{12} \diagdown\!\!\!\!\diagup \frac{3}{7}$$

35   36

Because 35 is less than 36, $\frac{5}{12}$ is less than $\frac{3}{7}$.

19 $\frac{3}{29}$ **is greater than** $\frac{1}{10}$ **and** $\frac{2}{21}$. Use cross-multiplication to compare the first two fractions.

$$\frac{1}{10} \diagdown\!\!\!\!\diagup \frac{2}{21}$$

21   20

Because 21 is greater than 20, $\frac{1}{10}$ is greater than $\frac{2}{21}$, so you can rule out $\frac{2}{21}$. Next, compare $\frac{1}{10}$ and $\frac{3}{29}$ by cross-multiplying.

$$\frac{1}{10} \diagdown\!\!\!\!\diagup \frac{3}{29}$$

29   30

Because 30 is greater than 29, $\frac{3}{29}$ is greater than $\frac{1}{10}$. Therefore, $\frac{3}{29}$ is the greatest of the three fractions.

20 $\frac{2}{7}$ **is less than** $\frac{1}{3}$, $\frac{4}{13}$, **and** $\frac{8}{25}$. Cross-multiply to compare the first two fractions.

$$\frac{1}{3} \diagdown\!\!\!\!\diagup \frac{2}{7}$$

7   6

Because 6 is less than 7, $\frac{2}{7}$ is less than $\frac{1}{3}$, so you can rule out $\frac{1}{3}$. Next, compare $\frac{2}{7}$ and $\frac{4}{13}$:

$$\frac{2}{7} \diagdown\!\!\!\!\diagup \frac{4}{13}$$

26   27

Because 26 is less than 27, $\frac{2}{7}$ is less than $\frac{4}{13}$, so you can rule out $\frac{4}{13}$.
Finally, compare $\frac{2}{7}$ and $\frac{8}{25}$:

$$\frac{2}{7} \diagdown\!\!\!\!\diagup \frac{8}{25}$$

50   56

Because 50 is less than 56, $\frac{2}{7}$ is less than $\frac{8}{25}$. Therefore, $\frac{2}{7}$ is the lowest of the four fractions.

21 **28.** To begin, set up a proportion based on the ratio of vegetables to fruit:

$$\frac{\text{Vegetables}}{\text{Fruit}} = \frac{4}{5}$$

Now, increase the terms of the fraction $\frac{4}{5}$ so that the number representing fruit becomes 35:

$$\frac{\text{Vegetables}}{\text{Fruit}} = \frac{4 \times 7}{5 \times 7} = \frac{28}{35}$$

Therefore, given that the farmers market has 35 varieties of fruit, it has 28 varieties of vegetables.

22 **63.** To begin, set up a proportion based on the ratio of sculptures to paintings:

$$\frac{\text{Sculptures}}{\text{Paintings}} = \frac{2}{7}$$

Now, increase the terms of the fraction $\frac{2}{7}$ so that the number representing sculptures becomes 18:

$$\frac{\text{Sculptures}}{\text{Paintings}} = \frac{2 \times 9}{7 \times 9} = \frac{18}{63}$$

Therefore, the gallery has 63 paintings.

23 **208.** To begin, set up a proportion based on the ratio of girls to boys:

$$\frac{\text{Girls}}{\text{Boys}} = \frac{7}{9}$$

Now, you want to increase the terms of the fraction $\frac{7}{9}$ so that the number representing boys becomes 117. (To do this, notice that $117 \div 9 = 13$, so $9 \times 13 = 117$):

$$\frac{\text{Girls}}{\text{Boys}} = \frac{7 \times 13}{9 \times 13} = \frac{91}{117}$$

Therefore, the camp has 91 girls and 117 boys, so the total number of children is 208.

24 **$825,000.** To begin, set up a proportion based on the ratio of state to municipal funding:

$$\frac{\text{State}}{\text{Municipal}} = \frac{3}{8}$$

Now, you want to increase the terms of the fraction $\frac{3}{8}$ so that the number representing municipal funding becomes 600,000. (To do this, notice that $600,000 \div 8 = 75,000$, so $8 \times 75,000 = 600,000$):

$$\frac{\text{State}}{\text{Municipal}} = \frac{3 \times 75,000}{8 \times 75,000} = \frac{225,000}{600,000}$$

Therefore, the town's budget includes $225,000 in state funding and $600,000 in municipal funding, so the total budget is $825,000.

Chapter 7

Fractions and the Big Four

. . . .

In This Chapter

▶ Multiplying and dividing fractions

▶ Knowing a variety of methods for adding and subtracting fractions

▶ Applying the Big Four operations to mixed numbers

. . . .

After you get the basics of fractions (which I cover in Chapter 6), you need to know how to apply the Big Four operations — adding, subtracting, multiplying, and dividing — to fractions and mixed numbers. In this chapter, I get you up to speed. First, I show you how to multiply and divide fractions — surprisingly, these two operations are the easiest to do. Next, I show you how to add fractions that have a common denominator (that is, fractions that have the same bottom number). After that, you discover a couple ways to add fractions that have different denominators. Then I repeat this process for subtracting fractions.

After that, I focus on mixed numbers. Again, I start with multiplication and division and then move on to the more difficult operations of addition and subtraction. At the end of this chapter, you should have a solid understanding of how to apply each of the Big Four operations to both fractions and mixed numbers.

Multiplying Fractions: A Straight Shot

Why can't everything in life be as easy as multiplying fractions? To multiply two fractions, just do the following:

▸ Multiply the two *numerators* (top numbers) to get the numerator of the answer.

▸ Multiply the two *denominators* (bottom numbers) to get the denominator of the answer.

When you multiply two proper fractions, the answer is always a proper fraction, so you won't have to change it to a mixed number, but you may have to reduce it. (See Chapter 6 for more on reducing fractions.)

Before you multiply, see whether you can cancel out common factors that appear in both the numerator and denominator. (This process is similar to reducing a fraction.) When you cancel out all common factors before you multiply, you get an answer that's already reduced to lowest terms.

Q. Multiply $\frac{2}{5}$ by $\frac{4}{9}$.

A. $\frac{8}{45}$. Multiply the two numerators (top numbers) to get the numerator of the answer. Then multiply the two denominators (bottom numbers) to get the denominator of the answer:

$$\frac{2}{5} \times \frac{4}{9} = \frac{(2 \times 4)}{(5 \times 9)} = \frac{8}{45}$$

In this case, you don't have to reduce the answer.

Q. Find $\frac{4}{7} \times \frac{5}{8}$.

A. $\frac{5}{14}$. Before you multiply, notice that the numerator 4 and the denominator 8 are both even. So, divide both of these numbers by 2 just as you would when reducing a fraction:

$$\frac{4}{7} \times \frac{5}{8} = \frac{2}{7} \times \frac{5}{4}$$

The numerator 2 and the denominator 4 are still even, so repeat the process:

$$= \frac{1}{7} \times \frac{5}{2}$$

At this point, neither numerator has a common factor with either denominator, so you're ready to multiply. Multiply the two numerators to get the numerator of the answer. Then multiply the two denominators to get the denominator of the answer:

$$\frac{(1 \times 5)}{(7 \times 2)} = \frac{5}{14}$$

Because you canceled all common factors before multiplying, this answer is in lowest terms.

1. Multiply $\frac{2}{3}$ by $\frac{7}{9}$.

Solve It

2. Find $\frac{3}{8} \times \frac{6}{11}$.

Solve It

3. Multiply $\frac{2}{9}$ by $\frac{3}{10}$.

Solve It

4. Figure out $\frac{9}{14} \times \frac{8}{15}$.

Solve It

Flipping for Fraction Division

Mathematicians didn't want to muck up fraction division by making you do something as complicated as actually *dividing*, so they devised a way to use multiplication instead. To divide one fraction by another fraction, change the problem to multiplication:

1. Change the division sign to a multiplication sign.

2. Change the *second* fraction to its reciprocal.

   Switch around the numerator (top number) and denominator (bottom number).

3. Solve the problem using fraction multiplication.

When dividing fractions, you may have to reduce your answer or change it from an improper fraction to a mixed number, as I show you in Chapter 6.

Q. Divide $\frac{5}{8}$ by $\frac{3}{7}$.

A. $1\frac{11}{24}$. Change the division to multiplication:

$$\frac{5}{8} \div \frac{3}{7} = \frac{5}{8} \times \frac{7}{3}$$

Solve the problem using fraction multiplication:

$$= \frac{(5 \times 7)}{(8 \times 3)} = \frac{35}{24}$$

The answer is an improper fraction (because the numerator is greater than the denominator), so change it to a mixed number. Divide the numerator by the denominator and put the remainder over the denominator:

$$= 1\frac{11}{24}$$

Q. Calculate $\frac{7}{10} \div \frac{2}{5}$.

A. $1\frac{3}{4}$. Change the division to multiplication:

$$\frac{7}{10} \div \frac{2}{5} = \frac{7}{10} \times \frac{5}{2}$$

Notice that you have a 5 in one of the numerators and a 10 in the other fraction's denominator, so you can cancel out the common factor, which is 5; that would change your problem to $\frac{7}{2} \times \frac{1}{2}$. Or you can simply do your calculations and reduce the fraction later, as I do here. Solve by multiplying these two fractions:

$$= \frac{(7 \times 5)}{(10 \times 2)} = \frac{35}{20}$$

This time, the numerator and denominator are both divisible by 5, so you can reduce them:

$$= \frac{7}{4}$$

Because the numerator is greater than the denominator, the fraction is improper, so change it to a mixed number:

$$1\frac{3}{4}$$

5. Divide $\frac{1}{4}$ by $\frac{6}{7}$.

Solve It

6. Find $\frac{3}{5} \div \frac{9}{10}$.

Solve It

7. Divide $\frac{8}{9}$ by $\frac{3}{10}$.

Solve It

8. Solve $\frac{14}{15} \div \frac{7}{12}$.

Solve It

Reaching the Common Denominator: Adding Fractions

In this section, I show you the easy stuff first, but then things get a little trickier. Adding fractions that have the same denominator (also called a *common denominator*) is super easy: Just add the numerators and keep the denominator the same. Sometimes you may have to reduce the answer to lowest terms or change it from an improper fraction to a mixed number.

Adding fractions that have different denominators takes a bit of work. Essentially, you need to increase the terms of one or both fractions so the denominators match before you can add. The easiest way to do this is by using a cross-multiplication trick that switches the terms of the fractions for you. Here's how it works:

1. **Cross-multiply the two fractions (as I show you in Chapter 6).**

   Multiply the numerator of the first fraction by the denominator of the second fraction and multiply the numerator of the second fraction by the denominator of the first.

2. **Build two fractions that have a common denominator.**

   Multiply the denominators of your two original fractions to get the new, common denominator. Create two new fractions by putting your results from Step 1 over this new denominator.

3. **Add the fractions from Step 2.**

   Add the numerators and leave the denominator the same.

When one denominator is a multiple of the other, you can use a quick trick to find a common denominator: Increase only the terms of the fraction with the lower denominator to make both denominators the same.

Q. Find $\frac{5}{8} + \frac{7}{8}$.

A. $1\frac{1}{2}$. The denominators are both 8, so add the numerators (5 and 7) to get the new numerator and keep the denominator the same:

$$\frac{5}{8} + \frac{7}{8} = \frac{(5+7)}{8} = \frac{12}{8}$$

The numerator is greater than the denominator, so the answer is an improper fraction. Change it to a mixed number and then reduce (as I show you in Chapter 6):

$$= 1\frac{2}{4} = 1\frac{1}{2}$$

Q. Add $\frac{3}{5}$ and $\frac{14}{15}$.

A. $1\frac{8}{15}$. The denominators are different, but because 15 is a multiple of 5, you can use the quick trick described earlier. Increase the terms of $\frac{3}{5}$ so that its denominator is 15. To do this, you need to multiply the numerator and the denominator by the same number. You have to multiply 5 times 3 to get 15 in the denominator, so you want to multiply the numerator by 3 as well:

$$\frac{3}{5} = \frac{(3 \times 3)}{(5 \times 3)} = \frac{9}{15}$$

Now both fractions have the same denominator, so add their numerators:

$$= \frac{9}{15} + \frac{14}{15} = \frac{(9+14)}{15} = \frac{23}{15}$$

The result is an improper fraction, so change this to a mixed number:

$$= 1\frac{8}{15}$$

Q. What is $\frac{1}{2} + \frac{2}{5}$?

A. $\frac{9}{10}$. Cross-multiply and add the results to find the numerator of the answer. Then multiply the denominators to find the denominator of the answer:

$$\frac{1}{2} + \frac{2}{5} = \frac{(5+4)}{10} = \frac{9}{10}$$

9. Add $\frac{7}{9}$ and $\frac{8}{9}$.

Solve It

10. Solve $\frac{3}{7} + \frac{4}{11}$.

Solve It

11. Find $\frac{5}{6} + \frac{7}{10}$.

Solve It

12. Add $\frac{8}{9} + \frac{17}{18}$.

Solve It

13. Find $\frac{12}{13} + \frac{9}{14}$.

Solve It

14. Add $\frac{9}{10}$ and $\frac{47}{50}$.

Solve It

15. Find the sum of $\frac{3}{17}$ and $\frac{10}{19}$.

Solve It

16. Add $\frac{3}{11} + \frac{5}{99}$.

Solve It

The Other Common Denominator: Subtracting Fractions

As with addition, subtracting fractions that have the same denominator (also called a *common denominator*) is very simple: Just subtract the second numerator from the first and keep the denominator the same. In some cases, you may have to reduce the answer to lowest terms.

Subtracting fractions that have different denominators takes a bit more work. You need to increase the terms of one or both fractions so both fractions have the same denominator. The easiest way to do this is to use cross-multiplication:

1. **Cross-multiply the two fractions (as I show you in Chapter 6) and create two fractions that have a common denominator.**

2. **Subtract the results from Step 1.**

When one denominator is a factor of the other, you can use a quick trick to find a common denominator: Increase only the terms of the fraction with the lower denominator to make both denominators the same.

Q. Find $\frac{5}{6} - \frac{1}{6}$.

A. $\frac{2}{3}$. The denominators are both 6, so subtract the numerators (5 and 1) to get the new numerator, and keep the denominator the same:

$$\frac{5}{6} - \frac{1}{6} = \frac{(5-1)}{6} = \frac{4}{6}$$

The numerator and denominator are both even numbers, so you can reduce the fraction by a factor of 2:

$$= \frac{2}{3}$$

Q. Find $\frac{6}{7} - \frac{17}{28}$.

A. $\frac{1}{4}$. The denominators are different, but because 28 is a multiple of 7, you can use the quick trick described earlier. Increase the terms of $\frac{6}{7}$ so that its denominator is 28; because $28 = 7 \times 4$, multiply both the numerator and denominator by 4:

$$\frac{6}{7} = \frac{(6 \times 4)}{(7 \times 4)} = \frac{24}{28}$$

Now both fractions have the same denominator, so subtract the numerators and keep the same denominator:

$$= \frac{24}{28} - \frac{17}{28} = \frac{(24-17)}{28} = \frac{7}{28}$$

Both the numerator and denominator are divisible by 7, so you can reduce this fraction by a factor of 7:

$$= \frac{1}{4}$$

17. Subtract $\frac{7}{10} - \frac{3}{10}$.

Solve It

18. Find $\frac{4}{5} - \frac{1}{3}$.

Solve It

19. Solve $\frac{5}{6} - \frac{7}{12}$.

Solve It

20. Subtract $\frac{10}{11} - \frac{4}{7}$.

Solve It

21. Solve $\frac{1}{4} - \frac{5}{22}$.

Solve It

22. Find $\frac{13}{15} - \frac{14}{45}$.

Solve It

23. Subtract $\frac{11}{12} - \frac{73}{96}$.

Solve It

24. What is $\frac{1}{999} - \frac{1}{1,000}$?

Solve It

Multiplying and Dividing Mixed Numbers

To multiply or divide two mixed numbers, convert both to improper fractions (as I show you in Chapter 6); then multiply or divide them just like any other fractions. At the end, you may have to reduce the result to lowest terms or convert the result back to a mixed number.

Q. What is $2\frac{1}{5} \times 3\frac{1}{4}$?

A. $7\frac{3}{20}$. First, convert both mixed numbers to improper fractions. Multiply the whole number by the denominator and add the numerator; then place your answer over the original denominator:

$$2\frac{1}{5} = \frac{(2 \times 5 + 1)}{5} = \frac{11}{5}$$
$$3\frac{1}{4} = \frac{(3 \times 4 + 1)}{4} = \frac{13}{4}$$

Now multiply the two fractions:

$$\frac{11}{5} \times \frac{13}{4} = \frac{(11 \times 13)}{5 \times 4} = \frac{143}{20}$$

Because the answer is an improper fraction, change it back to a mixed number:

$$\begin{array}{r} 7 \\ 20\overline{)143} \\ -140 \\ \hline 3 \end{array}$$

The final answer is $7\frac{3}{20}$.

Q. What is $3\frac{1}{2} \div 1\frac{1}{7}$?

A. $3\frac{1}{16}$. First, convert both mixed numbers to improper fractions:

$$3\frac{1}{2} = \frac{(3 \times 2 + 1)}{2} = \frac{7}{2}$$
$$1\frac{1}{7} = \frac{(1 \times 7 + 1)}{7} = \frac{8}{7}$$

Now divide the two fractions:

$$\frac{7}{2} \div \frac{8}{7} = \frac{7}{2} \times \frac{7}{8} = \frac{49}{16}$$

Because the answer is an improper fraction, convert it to a mixed number:

$$\begin{array}{r} 3 \\ 16\overline{)49} \\ -48 \\ \hline 1 \end{array}$$

The final answer is $3\frac{1}{16}$.

25. Multiply $2\frac{1}{3}$ by $1\frac{3}{7}$.

Solve It

26. Find $2\frac{2}{5} \times 1\frac{5}{6}$.

Solve It

27. Multiply $4\frac{4}{5}$ by $3\frac{1}{8}$.

Solve It

28. Calculate $4\frac{1}{2} \div 1\frac{5}{8}$.

Solve It

29. Divide $2\frac{1}{10}$ by $2\frac{1}{4}$.

Solve It

30. Find $1\frac{2}{7} \div 6\frac{3}{10}$.

Solve It

Carried Away: Adding Mixed Numbers

Adding mixed numbers isn't really more difficult than adding fractions. Here's how the process works:

1. **Add the fractional parts, reducing the answer if necessary.**

2. **If the answer you found in Step 1 is an improper fraction, change it to a mixed number, write down the fractional part, and *carry* the whole-number part to the whole-number column.**

3. **Add the whole-number parts (including any number you carried over).**

Q. Add $4\frac{1}{8} + 2\frac{3}{8}$.

A. $6\frac{1}{2}$. To start out, set up the problem in column form:

$$4\frac{1}{8}$$
$$+2\frac{3}{8}$$

Add the fractions and reduce the result:

$$\frac{1}{8} + \frac{3}{8} = \frac{4}{8} = \frac{1}{2}$$

Because this result is a proper fraction, you don't have to worry about carrying. Next, add the whole-number parts:

$$4 + 2 = 6$$

Here's how the problem looks in column form:

$$4\frac{1}{8}$$
$$+2\frac{3}{8}$$
$$\overline{6\frac{1}{2}}$$

Q. Add $5\frac{3}{4} + 4\frac{7}{9}$.

A. $10\frac{19}{36}$. To start out, set up the problem in column form:

$$5\frac{3}{4}$$
$$+4\frac{7}{9}$$

To add the fractional parts, change the two denominators to a common denominator using cross-multiplication. The new numerators are $3 \times 9 = 27$ and $7 \times 4 = 28$, and the new denominators are $4 \times 9 = 36$:

$$\frac{3}{4} \quad \frac{7}{9}$$
$$\downarrow \quad \downarrow$$
$$\frac{27}{36} \quad \frac{28}{36}$$

Now you can add:

$$\frac{27}{36} + \frac{28}{36} = \frac{55}{36}$$

Because this result is an improper fraction, change it to a mixed number.

$$= 1\frac{19}{36}$$

Carry the 1 from this mixed number into the whole-number column and add:

$$1 + 5 + 4 = 10$$

Here's how the problem looks in column form:

$$\overset{1}{}$$
$$5\frac{27}{36}$$
$$+4\frac{28}{36}$$
$$\overline{10\frac{19}{36}}$$

31. Add $3\frac{1}{5}$ and $4\frac{2}{5}$.

Solve It

32. Find $7\frac{1}{3}+1\frac{1}{6}$.

Solve It

33. Add $12\frac{4}{9}$ and $7\frac{8}{9}$.

Solve It

34. Find the sum of $5\frac{2}{3}$ and $9\frac{3}{5}$.

Solve It

35. Add $13\frac{6}{7}+2\frac{5}{14}$.

Solve It

36. Find $21\frac{9}{10}+38\frac{3}{4}$.

Solve It

Borrowing from the Whole: Subtracting Mixed Numbers

All right, I admit it: Most students find subtracting mixed numbers about as appealing as having their braces tightened. In this section, I attempt to make this process as painless as possible.

Subtracting mixed numbers is always easier when the denominators of the fractional parts are the same. When they're different, your first step is *always* to change them to fractions that have a common denominator. (In Chapter 6, I show you two ways of doing this — use whichever way works best.)

When the two mixed numbers have the same denominator, you're ready to subtract. To start out, I show you the simplest case. Here's how you subtract mixed numbers when the fractional part of the first number is *greater* than the fractional part of the second number:

1. **Find the difference of the fractional parts, reducing the result if necessary.**

2. **Find the difference of the whole-number parts.**

Subtracting mixed numbers gets a bit trickier when you need to borrow from the whole-number column into the fraction column. This is similar to borrowing in whole-number subtraction (for a refresher on borrowing, see Chapter 1).

Here's how you subtract mixed numbers when the fractional part of the first number is less than that of the second number:

1. **Borrow 1 from the whole-number column and add it to the fraction column, turning the top fraction into a mixed number.**

2. **Change this new mixed number into an improper fraction.**

3. **Use this result to subtract in the fraction column, reducing the result if necessary.**

4. **Perform the subtraction in the whole-number column.**

Q. Subtract $8\frac{4}{5} - 6\frac{3}{5}$.

A. $2\frac{1}{5}$. To start out, set up the problem in column form:

$$8\frac{4}{5}$$
$$-6\frac{3}{5}$$

The fractional parts already have a common denominator, so subtract:

$$\frac{4}{5} - \frac{3}{5} = \frac{1}{5}$$

Next, subtract the whole-number parts:

$$8 - 6 = 2$$

Here's how the completed problem looks:

$$8\frac{4}{5}$$
$$-6\frac{3}{5}$$
$$\overline{2\frac{1}{5}}$$

Q. Subtract $9\frac{1}{6} - 3\frac{5}{6}$.

A. $5\frac{1}{3}$. To start out, set up the problem in column form:

$$9\frac{1}{6}$$
$$-3\frac{5}{6}$$

The fractional parts already have a common denominator, so subtract. Notice that $\frac{1}{6}$ is less than $\frac{5}{6}$, so you need to borrow 1 from 9:

$$8\not{9}1\frac{1}{6}$$
$$-3\frac{5}{6}$$

Now convert the mixed number $1\frac{1}{6}$ into an improper fraction:

$$8\frac{7}{6}$$
$$-3\frac{5}{6}$$

At this point, you can subtract the fractional parts and reduce:

$$\frac{7}{6} - \frac{5}{6} = \frac{2}{6} = \frac{1}{3}$$

Next, subtract the whole-number parts:

$$8 - 3 = 5$$

Here's how the completed problem looks:

$$8\frac{7}{6}$$
$$-3\frac{5}{6}$$
$$5\frac{1}{3}$$

Q. Subtract $19\frac{4}{11} - 6\frac{3}{8}$.

A. $12\frac{87}{88}$. To start out, set up the problem in column form:

$$19\frac{4}{11}$$
$$-6\frac{3}{8}$$

The fractional parts have different denominators, so change them to a common denominator using cross-multiplication. The new numerators are $4 \times 8 = 32$ and $3 \times 11 = 33$, and the new denominators are $11 \times 8 = 88$:

$$\frac{4}{11} \quad \frac{3}{8}$$
$$\downarrow \quad \downarrow$$
$$\frac{32}{88} \quad \frac{33}{88}$$

Here's how the problem looks now:

$$19\frac{32}{88}$$
$$-6\frac{33}{88}$$

Because $\frac{32}{88}$ is less than $\frac{33}{88}$, you need to borrow before you can subtract:

$$18\not{19}1\frac{32}{88}$$
$$-6\frac{33}{88}$$

Now turn the mixed number $1\frac{32}{88}$ into an improper fraction:

$$18\frac{120}{88}$$
$$-6\frac{33}{88}$$

At this point, you can subtract the fractional parts and the whole-number parts:

$$18\frac{120}{88}$$
$$-6\frac{33}{88}$$
$$12\frac{87}{88}$$

37. Subtract $5\frac{7}{9} - 2\frac{4}{9}$.

Solve It

38. Find $9\frac{1}{8} - 7\frac{5}{8}$.

Solve It

39. Subtract $11\frac{3}{4} - 4\frac{2}{3}$.

Solve It

40. Figure out $16\frac{2}{5} - 8\frac{4}{9}$.

Solve It

Solutions to Fractions and the Big Four

The following are the answers to the practice questions presented in this chapter.

1 $\frac{2}{3} \times \frac{7}{9} = \frac{(2 \times 7)}{(3 \times 9)} = \frac{14}{27}$

2 $\frac{3}{8} \times \frac{6}{11} = \frac{9}{44}$. Multiply the numerator by the numerator and the denominator by the denominator:

$$\frac{3}{8} \times \frac{6}{11} = \frac{(3 \times 6)}{(8 \times 11)} = \frac{18}{88}$$

The numerator and denominator are both even, so both can be reduced by a factor of 2:

$$= \frac{9}{44}$$

3 $\frac{2}{9} \times \frac{3}{10} = \frac{1}{15}$. Begin by canceling out common factors. The numerator 2 and the denominator 10 are both even, so divide both by 2:

$$\frac{2}{9} \times \frac{3}{10} = \frac{1}{9} \times \frac{3}{5}$$

Next, the numerator 3 and the denominator 9 are both divisible by 3, so divide both by 3:

$$= \frac{1}{3} \times \frac{1}{5}$$

Now multiply straight across:

$$= \frac{(1 \times 1)}{(3 \times 5)} = \frac{1}{15}$$

Because you canceled out all common factors before multiplying, this answer is already reduced.

4 $\frac{9}{14} \times \frac{8}{15} = \frac{12}{35}$. Start by canceling out common factors. The numbers 14 and 8 are both divisible by 2, and 9 and 15 are divisible by 3:

$$\frac{9}{14} \times \frac{8}{15}$$
$$= \frac{9}{7} \times \frac{4}{15}$$
$$= \frac{3}{7} \times \frac{4}{5}$$

Now multiply:

$$= \frac{(3 \times 4)}{(7 \times 5)} = \frac{12}{35}$$

5 $\frac{1}{4} \div \frac{6}{7} = \frac{7}{24}$. First, change the problem to multiplication, multiplying by the reciprocal of the second fraction:

$$\frac{1}{4} \div \frac{6}{7} = \frac{1}{4} \times \frac{7}{6}$$

Now complete the problem using fraction multiplication:

$$= \frac{(1 \times 7)}{(4 \times 6)} = \frac{7}{24}$$

6 $\frac{3}{5} \div \frac{9}{10} = \frac{2}{3}$. Change the problem to multiplication, using the reciprocal of the second fraction:

$$\frac{3}{5} \div \frac{9}{10} = \frac{3}{5} \times \frac{10}{9}$$

Complete the problem using fraction multiplication:

$$= \frac{(3 \times 10)}{(5 \times 9)} = \frac{30}{45}$$

Both the numerator and denominator are divisible by 5, so reduce the fraction by this factor:

$$= \frac{6}{9}$$

They're still both divisible by 3, so reduce the fraction by this factor:

$$= \frac{2}{3}$$

7 $\frac{8}{9} \div \frac{3}{10} = 2\frac{26}{27}$. Change the problem to multiplication, using the reciprocal of the second fraction:

$$\frac{8}{9} \div \frac{3}{10} = \frac{8}{9} \times \frac{10}{3}$$

Complete the problem using fraction multiplication:

$$= \frac{(8 \times 10)}{(9 \times 3)} = \frac{80}{27}$$

The numerator is greater than the denominator, so change this improper fraction to a mixed number:

$$= 2\frac{26}{27}$$

8 $\frac{14}{15} \div \frac{7}{12} = 1\frac{3}{5}$. Change the problem to multiplication, using the reciprocal of the second fraction:

$$\frac{14}{15} \div \frac{7}{12} = \frac{14}{15} \times \frac{12}{7}$$

Complete the problem using fraction multiplication:

$$= \frac{(14 \times 12)}{(15 \times 7)} = \frac{168}{105}$$

The numerator is greater than the denominator, so change this improper fraction to a mixed number:

$$= 1\frac{63}{105}$$

Now the numerator and the denominator are both divisible by 3, so reduce the fractional part of this mixed number by a factor of 3:

$$= 1\frac{21}{35}$$

The numerator and denominator are now both divisible by 7, so reduce the fractional part by this factor:

$$= 1\frac{3}{5}$$

9 $\frac{7}{9} + \frac{8}{9} = 1\frac{2}{3}$. The denominators are the same, so add the numerators:

$$\frac{7}{9} + \frac{8}{9} = \frac{15}{9}$$

Both the numerator and denominator are divisible by 3, so reduce the fraction by 3:

$$= \frac{5}{3}$$

The result is an improper fraction, so convert it to a mixed number:

$$= 1\frac{2}{3}$$

10 $\frac{3}{7} + \frac{4}{11} = \frac{61}{77}$. The denominators are different, so change them to a common denominator by cross-multiplying. The new numerators are 3 × 11 = 33 and 4 × 7 = 28, and the new denominators are 7 × 11 = 77:

$$\frac{3}{7} \quad \frac{4}{11}$$
$$\downarrow \quad \downarrow$$
$$\frac{33}{77} \quad \frac{28}{77}$$

Now you can add:

$$\frac{33}{77} + \frac{28}{77} = \frac{61}{77}$$

11 $\frac{5}{6} + \frac{7}{10} = 1\frac{8}{15}$. The denominators are different, so change them to a common denominator by cross-multiplying. The new numerators are 5 × 10 = 50 and 7 × 6 = 42, and the new denominators are 6 × 10 = 60:

$$\frac{5}{6} \quad \frac{7}{10}$$
$$\downarrow \quad \downarrow$$
$$\frac{50}{60} \quad \frac{42}{60}$$

Now you can add:

$$\frac{50}{60} + \frac{42}{60} = \frac{92}{60}$$

Both the numerator and denominator are even, so reduce the fraction by 2:

$$= \frac{46}{30}$$

They're still both even, so reduce again by 2:

$$= \frac{23}{15}$$

The result is an improper fraction, so change it to a mixed number:

$$= 1\frac{8}{15}$$

12 $\frac{8}{9} + \frac{17}{18} = 1\frac{5}{6}$. The denominators are different, but 18 is a multiple of 9, so you can use the quick trick. Increase the terms of $\frac{8}{9}$ so that the denominator is 18, multiplying both the numerator and denominator by 2:

$$\frac{8}{9} = \frac{(8 \times 2)}{(9 \times 2)} = \frac{16}{18}$$

Now both fractions have the same denominator, so add the numerators and keep the same denominator:

$$= \frac{16}{18} + \frac{17}{18} = \frac{33}{18}$$

Both the numerator and denominator are divisible by 3, so reduce the fraction by 3:

$$= \frac{11}{6}$$

The result is an improper fraction, so change it to a mixed number:

$$= 1\frac{5}{6}$$

13 $\frac{12}{13} + \frac{9}{14} = 1\frac{103}{182}$. The denominators are different, so give the fractions a common denominator by cross-multiplying. The new numerators are 12 × 14 = 168 and 9 × 13 = 117, and the new denominators are 13 × 14 = 182:

$$\frac{12}{13} \quad \frac{9}{14}$$
$$\downarrow \quad \downarrow$$
$$\frac{168}{182} \quad \frac{117}{182}$$

Now you can add:

$$\frac{168}{182} + \frac{117}{182} = \frac{285}{182}$$

The result is an improper fraction, so change it to a mixed number:

$$= 1\frac{103}{182}$$

14 $\frac{9}{10} + \frac{47}{50} = 1\frac{21}{25}$. The denominators are different, but 50 is a multiple of 10, so you can use the quick trick. Increase the terms of $\frac{9}{10}$ so that the denominator is 50, multiplying both the numerator and denominator by 5:

$$\frac{9}{10} = \frac{(9 \times 5)}{(10 \times 5)} = \frac{45}{50}$$

Now both fractions have the same denominator, so add the numerators:

$$= \frac{45}{50} + \frac{47}{50} = \frac{92}{50}$$

Both the numerator and denominator are even, so reduce the fraction by 2:

$$= \frac{46}{25}$$

The result is an improper fraction, so change it to a mixed number:

$$= 1\frac{21}{25}$$

15 $\frac{3}{17} + \frac{10}{19} = \frac{227}{323}$. The denominators are different, so change them to a common denominator by cross-multiplying. The new numerators are 3 × 19 = 57 and 10 × 17 = 170, and the new denominators are 17 × 19 = 323:

$$\frac{3}{17} \quad \frac{10}{19}$$
$$\downarrow \quad \downarrow$$
$$\frac{57}{323} \quad \frac{170}{323}$$

Now you can add:

$$\frac{57}{323} + \frac{170}{323} = \frac{227}{323}$$

16 $\frac{3}{11} + \frac{5}{99} = \frac{\mathbf{32}}{\mathbf{99}}$. The denominators are different, but 99 is a multiple of 11, so you can use the quick trick. Increase the terms of $\frac{3}{11}$ so that the denominator is 99, multiplying both the numerator and denominator by 9:

$$\frac{3}{11} = \frac{(3 \times 9)}{(11 \times 9)} = \frac{27}{99}$$

Now you can add:

$$\frac{27}{99} + \frac{5}{99} = \frac{32}{99}$$

17 $\frac{7}{10} - \frac{3}{10} = \frac{\mathbf{2}}{\mathbf{5}}$. The denominators are the same, so subtract the numerators and keep the same denominator:

$$\frac{7}{10} - \frac{3}{10} = \frac{4}{10}$$

The numerator and denominator are both even, so reduce this fraction by a factor of 2:

$$= \frac{2}{5}$$

18 $\frac{4}{5} - \frac{1}{3} = \frac{\mathbf{7}}{\mathbf{15}}$. The denominators are different, so change them to a common denominator by cross-multiplying. The new numerators are 4 × 3 = 12 and 1 × 5 = 5, and the new denominators are 5 × 3 = 15:

$$\frac{4}{5} \quad \frac{1}{3}$$
$$\downarrow \quad \downarrow$$
$$\frac{12}{15} \quad \frac{5}{15}$$

Now you can subtract:

$$\frac{12}{15} - \frac{5}{15} = \frac{7}{15}$$

19 $\frac{5}{6} - \frac{7}{12} = \frac{\mathbf{1}}{\mathbf{4}}$. The denominators are different, but 6 is a factor of 12, so you can use the quick trick. Increase the terms of $\frac{5}{6}$ so that the denominator is 12, multiplying both the numerator and the denominator by 2:

$$\frac{5}{6} = \frac{(5 \times 2)}{(6 \times 2)} = \frac{10}{12}$$

Now the two fractions have the same denominator, so you can subtract easily:

$$\frac{10}{12} - \frac{7}{12} = \frac{3}{12}$$

The numerator and denominator are both divisible by 3, so reduce the fraction by a factor of 3:

$$= \frac{1}{4}$$

20 $\frac{10}{11} - \frac{4}{7} = \frac{26}{77}$. The denominators are different, so change them to a common denominator by cross-multiplying. The new numerators are $10 \times 7 = 70$ and $4 \times 11 = 44$, and the new denominators are $11 \times 7 = 77$:

$$\frac{10}{11} \quad \frac{4}{7}$$
$$\downarrow \quad \downarrow$$
$$\frac{70}{77} \quad \frac{44}{77}$$

Now you can subtract:

$$\frac{70}{77} - \frac{44}{77} = \frac{26}{77}$$

21 $\frac{1}{4} - \frac{5}{22} = \frac{1}{44}$. The denominators are different, so change them to a common denominator by cross-multiplying. The new numerators are $1 \times 22 = 22$ and $5 \times 4 = 20$, and the new denominators are $4 \times 22 = 88$:

$$\frac{1}{4} \quad \frac{5}{22}$$
$$\downarrow \quad \downarrow$$
$$\frac{22}{88} \quad \frac{20}{88}$$

Now you can subtract:

$$\frac{22}{88} - \frac{20}{88} = \frac{2}{88}$$

The numerator and denominator are both even, so reduce this fraction by a factor of 2:

$$= \frac{1}{44}$$

22 $\frac{13}{15} - \frac{14}{45} = \frac{5}{9}$. The denominators are different, but 45 is a multiple of 15, so you can use the quick trick. Increase the terms of $\frac{13}{15}$ so that the denominator is 45, multiplying the numerator and denominator by 3:

$$\frac{13}{15} = \frac{(13 \times 3)}{(15 \times 3)} = \frac{39}{45}$$

Now the two fractions have the same denominator, so you can subtract easily:

$$\frac{39}{45} - \frac{14}{45} = \frac{25}{45}$$

The numerator and denominator are both divisible by 5, so reduce the fraction by a factor of 5:

$$= \frac{5}{9}$$

23 $\frac{11}{12} - \frac{73}{96} = \mathbf{\frac{5}{32}}$. The denominators are different, but 96 is a multiple of 12, so you can use the quick trick. Increase the terms of $\frac{11}{12}$ so that the denominator is 96 by multiplying the numerator and denominator by 8:

$$\frac{11}{12} = \frac{(11 \times 8)}{(12 \times 8)} = \frac{88}{96}$$

Now you can subtract:

$$\frac{88}{96} - \frac{73}{96} = \frac{15}{96}$$

Both the numerator and denominator are divisible by 3, so reduce the fraction by a factor of 3:

$$= \frac{5}{32}$$

24 $\frac{1}{999} - \frac{1}{1,000} = \mathbf{\frac{1}{999,000}}$. The denominators are different, so change them to a common denominator by cross-multiplying:

$$\frac{1}{999} \qquad \frac{1}{1,000}$$
$$\downarrow \qquad \downarrow$$
$$\frac{1,000}{999,000} \qquad \frac{999}{999,000}$$

Now you can subtract:

$$\frac{1,000}{999,000} - \frac{999}{999,000} = \frac{1}{999,000}$$

25 $2\frac{1}{3} \times 1\frac{3}{7} = \mathbf{3\frac{1}{3}}$. Change both mixed numbers to improper fractions:

$$2\frac{1}{3} = \frac{(2 \times 3 + 1)}{3} = \frac{7}{3}$$
$$1\frac{3}{7} = \frac{(1 \times 7 + 3)}{7} = \frac{10}{7}$$

Set up the multiplication:

$$\frac{7}{3} \times \frac{10}{7}$$

Before you multiply, you can cancel out 7s in the numerator and denominator:

$$= \frac{1}{3} \times \frac{10}{1} = \frac{10}{3}$$

Because the answer is an improper fraction, change it to a mixed number:

$$\begin{array}{r} 3 \\ 3\overline{)10} \\ -9 \\ \hline 1 \end{array}$$

So, the final is $3\frac{1}{3}$.

26 $2\frac{2}{5} \times 1\frac{5}{6} = 4\frac{2}{5}$. Change both mixed numbers to improper fractions:

$$2\frac{2}{5} = \frac{(2 \times 5 + 2)}{5} = \frac{12}{5}$$

$$1\frac{5}{6} = \frac{(1 \times 6 + 5)}{6} = \frac{11}{6}$$

Set up the multiplication:

$$\frac{12}{5} \times \frac{11}{6}$$

Before you multiply, you can cancel out 6s in the numerator and denominator:

$$= \frac{2}{5} \times \frac{11}{1} = \frac{22}{5}$$

Because the answer is an improper fraction, change it to a mixed number:

$$\begin{array}{r} 4 \\ 5{\overline{)22}} \\ -20 \\ \hline 2 \end{array}$$

The final answer is $4\frac{2}{5}$.

27 $4\frac{4}{5} \times 3\frac{1}{8} = \mathbf{15}$. Change both mixed numbers to improper fractions:

$$4\frac{4}{5} = \frac{(4 \times 5 + 4)}{5} = \frac{24}{5}$$

$$3\frac{1}{8} = \frac{(3 \times 8 + 1)}{8} = \frac{25}{8}$$

Set up the multiplication:

$$\frac{24}{5} \times \frac{25}{8}$$

Before you multiply, cancel out both 5s and 8s in the numerator and denominator:

$$= \frac{24}{1} \times \frac{5}{8} = \frac{3}{1} \times \frac{5}{1} = 15$$

28 $4\frac{1}{2} \div 1\frac{5}{8} = \mathbf{2\frac{10}{13}}$. Change both mixed numbers to improper fractions:

$$4\frac{1}{2} = \frac{(4 \times 2 + 1)}{2} = \frac{9}{2}$$

$$1\frac{5}{8} = \frac{(1 \times 8 + 5)}{8} = \frac{13}{8}$$

Set up the division:

$$\frac{9}{2} \div \frac{13}{8}$$

Change the problem to multiplication using the reciprocal of the second fraction:

$$= \frac{9}{2} \times \frac{8}{13}$$

Cancel out a factor of 2 and multiply:

$$= \frac{9}{1} \times \frac{4}{13} = \frac{36}{13}$$

Because the answer is an improper fraction, change it to a mixed number:

$$= 2\frac{10}{13}$$

29 $2\frac{1}{10} \div 2\frac{1}{4} = \frac{\mathbf{14}}{\mathbf{15}}$. Change both mixed numbers to improper fractions:

$$2\frac{1}{10} = \frac{(2 \times 10 + 1)}{10} = \frac{21}{10}$$
$$2\frac{1}{4} = \frac{(2 \times 4 + 1)}{4} = \frac{9}{4}$$

Set up the division:

$$\frac{21}{10} \div \frac{9}{4}$$

Change the problem to multiplication using the reciprocal of the second fraction:

$$= \frac{21}{10} \times \frac{4}{9}$$

Before you multiply, cancel 2s and 3s from the numerator and denominator:

$$= \frac{21}{5} \times \frac{2}{9} = \frac{7}{5} \times \frac{2}{3} = \frac{14}{15}$$

30 $1\frac{2}{7} \div 6\frac{3}{10} = \frac{\mathbf{10}}{\mathbf{49}}$. Change both mixed numbers to improper fractions:

$$1\frac{2}{7} = \frac{(1 \times 7 + 2)}{7} = \frac{9}{7}$$
$$6\frac{3}{10} = \frac{(6 \times 10 + 3)}{10} = \frac{63}{10}$$

Set up the division:

$$\frac{9}{7} \div \frac{63}{10}$$

Change the problem to multiplication using the reciprocal of the second fraction:

$$= \frac{9}{7} \times \frac{10}{63}$$

Before you multiply, cancel 9s from the numerator and denominator:

$$= \frac{1}{7} \times \frac{10}{7} = \frac{10}{49}$$

31 $3\frac{1}{5} + 4\frac{2}{5} = \mathbf{7\frac{3}{5}}$. Set up the problem in column form:

$$3\frac{1}{5}$$
$$+4\frac{2}{5}$$

Add the fractional parts:

$$\frac{1}{5} + \frac{2}{5} = \frac{3}{5}$$

Because this result is a proper fraction, you don't have to worry about carrying. Next, add the whole-number parts:

$$3 + 4 = 7$$

Here's how the completed problem looks:

$$3\frac{1}{5}$$
$$+4\frac{2}{5}$$
$$\overline{7\frac{3}{5}}$$

32 $7\frac{1}{3} + 1\frac{1}{6} = \mathbf{8\frac{1}{2}}$. To start out, set up the problem in column form:

$$7\frac{1}{3}$$
$$+1\frac{1}{6}$$

Next, add the fractional parts. The denominators are different, but 3 is a factor of 6, so you can use the quick trick. Increase the terms of $\frac{1}{3}$ so that the denominator is 6 by multiplying the numerator and denominator by 2:

$$\frac{1}{3} = \frac{2}{6}$$

Now you can add and reduce the result:

$$\frac{2}{6} + \frac{1}{6} = \frac{3}{6} = \frac{1}{2}$$

Because this result is a proper fraction, you don't have to worry about carrying. Next, add the whole-number parts:

$$7 + 1 = 8$$

Here's how the completed problem looks:

$$7\frac{2}{6}$$
$$+1\frac{1}{6}$$
$$\overline{8\frac{1}{2}}$$

33 $12\frac{4}{9} + 7\frac{8}{9} = 20\frac{1}{3}$. Set up the problem in column form:

$$12\frac{4}{9}$$

$$+7\frac{8}{9}$$

Add the fractional parts and reduce the result:

$$\frac{4}{9} + \frac{8}{9} = \frac{12}{9} = \frac{4}{3}$$

Because this result is an improper fraction, convert it to a mixed number:

$$= 1\frac{1}{3}$$

Carry the 1 from this mixed number into the whole-number column and add:

$$1 + 12 + 7 = 20$$

Here's how the completed problem looks:

$$12\overset{1}{}\frac{4}{9}$$

$$+7\frac{8}{9}$$

$$20\frac{1}{3}$$

34 $5\frac{2}{3} + 9\frac{3}{5} = 15\frac{4}{15}$. Set up the problem in column form:

$$5\frac{2}{3}$$

$$+9\frac{3}{5}$$

Start by adding the fractional parts. Because the denominators are different, change them to a common denominator by cross-multiplying. The new numerators are $2 \times 5 = 10$ and $3 \times 3 = 9$, and the new denominators are $3 \times 5 = 15$:

$$\frac{2}{3} \quad \frac{3}{5}$$

$$\downarrow \quad \downarrow$$

$$\frac{10}{15} \quad \frac{9}{15}$$

Now you can add:

$$\frac{10}{15} + \frac{9}{15} = \frac{19}{15}$$

Because this result is an improper fraction, convert it to a mixed number:

$$= 1\frac{4}{15}$$

Carry the 1 from this mixed number into the whole-number column and add:

$$1 + 5 + 9 = 15$$

Here's how the completed problem looks:

$$5\overset{1}{}\frac{10}{15}$$
$$+9\frac{9}{15}$$
$$\overline{}$$
$$15\frac{4}{15}$$

35 $13\frac{6}{7} + 2\frac{5}{14} = 16\frac{3}{14}$. Set up the problem in column form:

$$13\frac{6}{7}$$
$$+2\frac{5}{14}$$
$$\overline{}$$

Begin by adding the fractional parts. Because the denominator 7 is a factor of the denominator 14, you can use the quick trick. Increase the terms of $\frac{6}{7}$ so that the denominator is 14 by multiplying the numerator and denominator by 2:

$$\frac{6}{7} = \frac{12}{14}$$

Now you can add:

$$\frac{12}{14} + \frac{5}{14} = \frac{17}{14}$$

Because this result is an improper fraction, convert it to a mixed number:

$$= 1\frac{3}{14}$$

Carry the 1 from this mixed number into the whole-number column and add:

$$1 + 13 + 2 = 16$$

Here's how the completed problem looks:

$$13\overset{1}{}\frac{12}{14}$$
$$+2\frac{5}{14}$$
$$\overline{}$$
$$16\frac{3}{14}$$

36 $21\frac{9}{10} + 38\frac{3}{4} = 60\frac{13}{20}$. Set up the problem in column form:

$$21\frac{9}{10}$$
$$+38\frac{3}{4}$$
$$\overline{}$$

To add the fractional parts, change the denominators to a common denominator by using cross-multiplication. The new numerators are $9 \times 4 = 36$ and $3 \times 10 = 30$, and the new denominators are $10 \times 4 = 40$:

$$\frac{9}{10} \quad \frac{3}{4}$$
$$\downarrow \quad \downarrow$$
$$\frac{36}{40} \quad \frac{30}{40}$$

Now you can add:

$$\frac{36}{40} + \frac{30}{40} = \frac{66}{40}$$

The numerator and denominator are both even, so reduce this fraction by a factor of 2:

$$= \frac{33}{20}$$

Because this result is an improper fraction, convert it to a mixed number:

$$= 1\frac{13}{20}$$

Carry the 1 from this mixed number into the whole-number column and add:

$$1 + 21 + 38 = 60$$

Here's how the completed problem looks:

$$21\overset{1}{}\frac{36}{40}$$
$$+38\frac{30}{40}$$
$$\overline{60\frac{13}{20}}$$

37 $5\frac{7}{9} - 2\frac{4}{9} = \mathbf{3\frac{1}{3}}$. Set up the problem in column form:

$$5\frac{7}{9}$$
$$-2\frac{4}{9}$$
$$\overline{}$$

Subtract the fractional parts and reduce:

$$\frac{7}{9} - \frac{4}{9} = \frac{3}{9} = \frac{1}{3}$$

Subtract the whole-number parts:

$$5 - 2 = 3$$

Here's how the completed problem looks:

$$5\frac{7}{9}$$
$$-2\frac{4}{9}$$
$$\overline{3\frac{1}{3}}$$

38 $9\frac{1}{8} - 7\frac{5}{8} = 1\frac{1}{2}$. Set up the problem in column form:

$$9\frac{1}{8}$$
$$-7\frac{5}{8}$$

The first fraction $\left(\frac{1}{8}\right)$ is less than the second fraction $\left(\frac{5}{8}\right)$, so you need to borrow 1 from 9 before you can subtract:

$$8\cancel{9}1\frac{1}{8}$$
$$-7\frac{5}{8}$$

Change the mixed number $1\frac{1}{8}$ to an improper fraction:

$$8\frac{9}{8}$$
$$-7\frac{5}{8}$$

Now you can subtract the fractional parts and reduce:

$$\frac{9}{8} - \frac{5}{8} = \frac{4}{8} = \frac{1}{2}$$

Subtract the whole-number parts:

$$8 - 7 = 1$$

Here's how the completed problem looks:

$$8\frac{9}{8}$$
$$-7\frac{5}{8}$$
$$1\frac{1}{2}$$

39 $11\frac{3}{4} - 4\frac{2}{3} = 7\frac{1}{12}$. Set up the problem in column form:

$$11\frac{3}{4}$$
$$-4\frac{2}{3}$$

The denominators are different, so get a common denominator using cross-multiplication. The new numerators are $3 \times 3 = 9$ and $2 \times 4 = 8$, and the new denominators are $4 \times 3 = 12$:

$$\frac{3}{4} \quad \frac{2}{3}$$
$$\downarrow \quad \downarrow$$
$$\frac{9}{12} \quad \frac{8}{12}$$

Because $\frac{9}{12}$ is greater than $\frac{8}{12}$, you don't need to borrow before you can subtract fractions:

$$11\frac{9}{12}$$
$$\underline{-4\frac{8}{12}}$$
$$7\frac{1}{12}$$

40 $16\frac{2}{5}-8\frac{4}{9}=\mathbf{7\frac{43}{45}}$. To start out, set up the problem in column form:

$$16\frac{2}{5}$$
$$\underline{-8\frac{4}{9}}$$

The denominators are different, so find a common denominator using cross-multiplication. The new numerators are $2 \times 9 = 18$ and $4 \times 5 = 20$, and the new denominators are $5 \times 9 = 45$:

$$\frac{2}{5} \quad \frac{4}{9}$$
$$\downarrow \quad \downarrow$$
$$\frac{18}{45} \quad \frac{20}{45}$$

Because $\frac{18}{45}$ is less than $\frac{20}{45}$, you need to borrow 1 from 16 before you can subtract fractions:

$$15\cancel{16}1\frac{18}{45}$$
$$\underline{-8\frac{20}{45}}$$

Change the mixed number $1\frac{18}{45}$ to an improper fraction:

$$15\frac{63}{45}$$
$$\underline{-8\frac{20}{45}}$$

Now you can subtract:

$$15\frac{63}{45}$$
$$\underline{-8\frac{20}{45}}$$
$$7\frac{43}{45}$$

Chapter 8

Getting to the Point with Decimals

....

....

Decimals, like fractions, are a way to represent parts of the whole — that is, positive numbers less than 1. You can use a decimal to represent any fractional amount. Decimals are commonly used for amounts of money, so you're probably familiar with the decimal point (.), which indicates an amount smaller than one dollar.

In this chapter, I first get you up to speed on some basic facts about decimals. Then I show you how to do basic conversions between fractions and decimals. After that, you find out how to apply the Big Four operations (adding, subtracting, multiplying, and dividing) to decimals. To end the chapter, I make sure you understand how to convert any fraction to a decimal and any decimal to a fraction. This includes filling you in on the difference between a *terminating decimal* (a decimal with a limited number of digits) and a *repeating decimal* (a decimal that repeats a pattern of digits endlessly).

Getting in Place: Basic Decimal Stuff

Decimals are easier to work with than fractions because they resemble whole numbers much more than fractions do. Decimals use place value in a similar way to whole numbers. In a decimal, however, each place represents a part of the whole. Take a look at the following chart.

Thousands	Hundreds	Tens	Ones	Decimal Point	Tenths	Hundredths	Thousandths
				.			

Notice that the name of each decimal place, from left to right, is linked with the name of a different fraction: tenths $\left(\frac{1}{10}\right)$, hundredths $\left(\frac{1}{100}\right)$, thousandths $\left(\frac{1}{1,000}\right)$, and so on.

You can use this chart to *expand* a decimal out as a sum. Expanding a decimal gives you a better sense of how that decimal is put together. For example, 12.011 is equal to

$$10 + 2 + \frac{0}{10} + \frac{1}{100} + \frac{1}{1,000}.$$

In a decimal, any zero to the right of both the decimal point and every nonzero digit is called a trailing zero. For example, in the decimal 0.070, the last zero is a trailing zero. You can safely drop this zero without changing the value of the decimal. However, the first 0 after the decimal point — which is trapped between the decimal point and a nonzero number — is a placeholding zero, which you can't drop.

You can express any whole number as a decimal simply by attaching a decimal point and a trailing zero to the end of it. For example,

$$7 = 7.0 \qquad 12 = 12.0 \qquad 1,568 = 1,568.0$$

In Chapter 2, I introduce the powers of ten: 1, 10, 100, 1,000, and so forth. Moving the decimal point to the *right* is the same as multiplying that decimal by a power of 10. For example,

- ✔ Moving the decimal point one place to the right is the same as multiplying by 10.
- ✔ Moving the decimal point two places to the right is the same as multiplying by 100.
- ✔ Moving the decimal point three places to the right is the same as multiplying by 1,000.

Similarly, moving the decimal point to the *left* is the same as dividing that decimal by a power of 10. For example,

- ✔ Moving the decimal point one place to the left is the same as dividing by 10.
- ✔ Moving the decimal point two places to the left is the same as dividing by 100.
- ✔ Moving the decimal point three places to the left is the same as dividing by 1,000.

To multiply a decimal by any power of 10, count the number of zeros and move the decimal point that many places to the right. To divide a decimal by any power of 10, count the number of zeros and move the decimal point that many places to the left.

Rounding decimals is similar to rounding whole numbers (if you need a refresher, see Chapter 1). Generally speaking, to round a number to a given decimal place, focus on that decimal place and the place to its immediate right; then round as you would with whole numbers:

- ✔ **Rounding down:** If the digit on the right is 0, 1, 2, 3, or 4, drop this digit and every digit to its right.
- ✔ **Rounding up:** If the digit on the right is 5, 6, 7, 8, or 9, add 1 to the digit you're rounding to and then drop every digit to its right.

When rounding, people often refer to the first three decimal places in two different ways — by the number of the decimal place and by the name:

- ✔ Rounding to *one decimal place* is the same as rounding *to the nearest tenth*.
- ✔ Rounding to *two decimal places* is the same as rounding *to the nearest hundredth*.
- ✔ Rounding to *three decimal places* is the same as rounding *to the nearest thousandth*.

When rounding to four or more decimal places, the names get longer, so they're usually not used.

Q. Expand the decimal 7,358.293.

A. $7,358.293 = 7,000 + 300 + 50 + 8 + \frac{2}{10} + \frac{9}{100} + \frac{3}{1,000}$.

Q. Simplify the decimal 0400.0600 by removing all leading and trailing zeros, without removing placeholding zeros.

A. **400.06.** The first zero is a leading zero because it appears to the left of all nonzero digits. The last two zeros are trailing zeros because they appear to the right of all nonzero digits. The remaining three zeros are placeholding zeros.

Q. Multiply 3.458 × 100.

A. **345.8.** The number 100 has two zeros, so to multiply by 100, move the decimal point two places to the right.

Q. Divide 29.81 ÷ 10,000.

A. **0.002981.** The number 10,000 has four zeros, so to divide by 10,000, move the decimal point four places to the left.

1. Expand the following decimals:

  a. 2.7

  b. 31.4

  c. 86.52

  d. 103.759

  e. 1,040.0005

  f. 16,821.1384

Solve It

2. Simplify each of the following decimals by removing all leading and trailing zeros whenever possible, without removing placeholding zeros:

  a. 5.80

  b. 7.030

  c. 90.0400

  d. 9,000.005

  e. 0108.0060

  f. 00100.0102000

Solve It

3. Do the following decimal multiplication and division problems by moving the decimal point the correct number of places:

a. 7.32×10

b. 9.04×100

c. $51.6 \times 100{,}000$

d. $2.786 \div 1{,}000$

e. $943.812 \div 1{,}000{,}000$

Solve It

4. Round each of the following decimals to the number of places indicated:

a. Round 4.777 to one decimal place.

b. Round 52.305 to the nearest tenth.

c. Round 191.2839 to two decimal places.

d. Round 99.995 to the nearest hundredth.

e. Round 0.00791 to three decimal places.

f. Round 909.9996 to the nearest thousandth.

Solve It

Simple Decimal-Fraction Conversions

Some conversions between decimals and fractions are easy to do. The conversions in Table 8-1 are all so common that they're worth memorizing. You can also use these to convert some decimals greater than one to mixed numbers, and vice versa.

Table 8-1	Equivalent Decimals and Fractions			
Tenths	**Eighths**	**Fifths**	**Quarters**	**Half**
$0.1 = \frac{1}{10}$	$0.125 = \frac{1}{8}$			
		$0.2 = \frac{1}{5}$	$0.25 = \frac{1}{4}$	
$0.3 = \frac{3}{10}$	$0.375 = \frac{3}{8}$			
		$0.4 = \frac{2}{5}$		
				$0.5 = \frac{1}{2}$
	$0.625 = \frac{5}{8}$	$0.6 = \frac{3}{5}$		
$0.7 = \frac{7}{10}$			$0.75 = \frac{3}{4}$	
	$0.875 = \frac{7}{8}$	$0.8 = \frac{4}{5}$		
$0.9 = \frac{9}{10}$				

EXAMPLE

Q. Convert 13.7 to a mixed number.

A. $13\frac{7}{10}$. The whole-number part of the decimal (13) becomes the whole-number part of the mixed number. Use the conversion chart to change the rest of the decimal (0.7) to a fraction.

Q. Change $9\frac{4}{5}$ to a decimal.

A. 9.8. The whole-number part of the mixed number (9) becomes the whole-number part of the decimal. Use the conversion chart to change the fractional part of the mixed number $\left(\frac{4}{5}\right)$ to a decimal.

5. Convert the following decimals into fractions:

a. 0.7

b. 0.4

c. 0.25

d. 0.125

e. 0.1

f. 0.75

Solve It

6. Change these fractions to decimals:

a. $\frac{9}{10}$

b. $\frac{2}{5}$

c. $\frac{3}{4}$

d. $\frac{3}{8}$

e. $\frac{7}{8}$

f. $\frac{1}{2}$

Solve It

7. Change these decimals to mixed numbers:

a. 1.6

b. 3.3

c. 14.5

d. 20.75

e. 100.625

f. 375.375

Solve It

8. Change these mixed numbers to decimals:

a. $1\frac{1}{5}$

b. $2\frac{1}{10}$

c. $3\frac{1}{2}$

d. $5\frac{1}{4}$

e. $7\frac{1}{8}$

f. $12\frac{5}{8}$

Solve It

New Lineup: Adding and Subtracting Decimals

You shouldn't lose much sleep at night worrying about adding and subtracting decimals, because it's nearly as easy as adding and subtracting whole numbers. Simply line up the decimal points and then add or subtract just as you would with whole numbers. The decimal point drops straight down in your answer.

To avoid mistakes (and to make your teacher happy), make sure your columns are neat. If you find it helpful, fill out the columns on the right with trailing zeros so that all numbers have the same number of decimal places. You may need to add these trailing zeros if you're subtracting a decimal from a number that has fewer decimal places.

Q. Add the following decimals: 321.81 + 24.5 + 0.006 = ?

A. **346.316.** Place the decimal numbers in a column (as you would for column addition) with the decimal points lined up:

$$321.810$$
$$24.500$$
$$\underline{+0.006}$$

Notice that the decimal point in the answer lines up with the others. As you can see, I've also filled out the columns with trailing zeros. This is optional, but do it if it helps you to see how the columns line up.

Now add as you would when adding whole numbers, carrying when necessary (see Chapter 2 for more on carrying in addition):

$$\overset{1}{3}21.810$$
$$24.500$$
$$\underline{+0.006}$$
$$346.316$$

Q. Subtract the following decimals: 978.245 − 29.03 = ?

A. **949.215.** Place the decimals one on top of the other with the decimal points lined up, dropping the decimal point straight down in the answer:

$$978.245$$
$$\underline{-29.030}$$

Now subtract as you would when subtracting whole numbers, borrowing when necessary (see Chapter 2 for more on borrowing in subtraction):

$$9\cancel{7}\,\overset{1}{8}.245$$
$$\underline{-29.030}$$
$$949.215$$

9. Add these decimals: 17.4 + 2.18 = ?

Solve It

10. Compute the following decimal addition:
0.0098 + 10.101 + 0.07 + 33 =

Solve It

11. Add the following decimals: 1,000.001 + 75 + 0.03 + 800.2 = ?

Solve It

12. Subtract these decimals: 0.748 – 0.23 = ?

Solve It

13. Compute the following: 674.9 – 5.0001.

Solve It

14. Find the solution to this decimal subtraction problem: 100.009 – 0.68 = ?

Solve It

Counting Decimal Places: Multiplying Decimals

To multiply two decimals, don't worry about lining up the decimal points. In fact, to start out, ignore the decimal points. Here's how the multiplication works:

1. **Perform the multiplication just as you would for whole numbers.**

2. **When you're done, count the number of digits to the right of the decimal point in each factor and add the results.**

3. **Place the decimal point in your answer so that your answer has the same number of digits after the decimal point.**

 Even if the last digit in the answer is 0, you still need to count this as a digit when placing the decimal point in a multiplication problem. After the decimal point is in place, however, you can drop trailing zeros.

 Q. Multiply the following decimals:
74.2 × 0.35 = ?

A. **25.97.** Ignoring the decimal points, perform the multiplication just as you would for whole numbers:

$$74.2$$
$$\underline{\times 0.35}$$
$$3710$$
$$\underline{+22260}$$
$$25970$$

At this point, you're ready to find out where the decimal point goes in the

answer. Count the number of decimal places in the two factors (74.2 and 0.35), add these two numbers together (1 + 2 = 3), and place the decimal point in the answer so that it has three digits to the right of the decimal point:

$$74.2 \leftarrow 1 \text{ digit after decimal}$$
$$\underline{\times 0.35} \leftarrow 2 \text{ digits after decimal}$$
$$3710$$
$$\underline{+22260}$$
$$25.970 \leftarrow 1+2 = 3 \text{ digits after decimal}$$

---

15. Multiply these decimals: 0.635 × 0.42 = ?

Solve It

16. Perform the following decimal multiplication: 0.675 × 34.8 = ?

Solve It

17. Solve the following multiplication problem: $943 \times 0.0012 = ?$

Solve It

18. Find the solution to this decimal multiplication: $1.006 \times 0.0807 = ?$

Solve It

Points on the Move: Dividing Decimals

Dividing decimals is similar to dividing whole numbers, except you have to handle the decimal point before you start dividing. Here's how to divide decimals step by step:

1. **Move the decimal point in the divisor and dividend.**

   Turn the *divisor* (the number you're dividing by) into a whole number by moving the decimal point all the way to the right. At the same time, move the decimal point in the *dividend* (the number you're dividing) the same number of places to the right.

2. **Place a decimal point in the *quotient* (the answer) directly above where the decimal point now appears in the dividend.**

3. **Divide as usual, being careful to line up the quotient properly so that the decimal point falls into place.**

   Line up each digit in the quotient just over the last digit in the dividend used in that cycle. Flip to Chapter 1 if you need a refresher on long division.

As with whole-number division, sometimes decimal division doesn't work out evenly at the end. With decimals, however, you never write a remainder. Instead, attach enough trailing zeros to round the quotient to a certain number of decimal places. The digit to the right of the digit you're rounding to tells you whether to round up or down, so you always have to figure out the division to one extra place (see "Getting in Place: Basic Decimal Stuff" earlier in this chapter for more on how to round decimals). See the following chart:

To Round a Decimal to	*Fill Out the Dividend with Trailing Zeros to*
A whole number	One decimal place
One decimal place	Two decimal places
Two decimal places	Three decimal places

Q. Divide the following: $9.152 \div 0.8 = ?$

A. **11.44.** To start out, write the problem as usual:

$$0.8\overline{)9.152}$$

Turn 0.8 into the whole number 8 by moving the decimal point one place to the right. At the same time, move the decimal point in 9.1526 one place to the right. Put your decimal point in the quotient directly above where it falls in 91.25:

$$8.\overline{)91.52}$$

Now you're ready to divide. Just be careful to line up the quotient properly so that the decimal point falls into place.

$$
\begin{array}{r}
11.44 \\
8.\overline{)91.52} \\
-8 \\
\hline
11 \\
-8 \\
\hline
35 \\
-32 \\
\hline
32 \\
-32 \\
\hline
0
\end{array}
$$

Q. Divide the following: $21.9 \div 0.015 = ?$

A. **1,460.** Set up the problem as usual:

$$0.015\overline{)21.900}$$

Notice that I attach two trailing zeros to the dividend. I do this because I need to move the decimal points in each number three places to the right. Again, place the decimal point in the quotient directly above where it now appears in the dividend, 21900:

$$15.\overline{)21900.}$$

Now you're ready to divide. Line up the quotient carefully so the decimal point falls into place:

$$
\begin{array}{r}
1460. \\
15.\overline{)21900.} \\
-15 \\
\hline
69 \\
-60 \\
\hline
90 \\
-90 \\
\hline
0
\end{array}
$$

Even though the division comes out even after you write the digit 6 in the quotient, you still need to add a placeholding zero so that the decimal point appears in the correct place.

19. Divide these two decimals: 9.345 ÷ 0.05 = ?

Solve It

20. Solve the following division: 3.15 ÷ .021 = ?

Solve It

21. Perform the following decimal division, rounding to one decimal place: 6.7 ÷ 10.1.

Solve It

22. Find the solution, rounding to the nearest hundredth: 9.13 ÷ 4.25.

Solve It

Decimals to Fractions

Some conversions from very common decimals to fractions are easy (see "Simple Decimal-Fraction Conversions" earlier in this chapter). In other cases, you have to do a bit more work. Here's how to change any decimal into a fraction:

1. **Create a "fraction" with the decimal in the numerator and 1.0 in the denominator.**

   This isn't really a fraction, because a fraction always has whole numbers in both the numerator and denominator, but you turn it into a fraction in Step 2.

   When converting a decimal that's greater than 1 to a fraction, separate out the whole-number part of the decimal before you begin; work only with the decimal part. The resulting fraction is a mixed number.

2. **Move the decimal point in the numerator enough places to the right to turn the numerator into a whole number, and move the decimal point in the denominator the same number of places.**

3. **Drop the decimal points and any trailing zeros.**

4. **Reduce the fraction to lowest terms if necessary.**

   See Chapter 6 for info on reducing fractions.

A quick way to make a fraction out of a decimal is to use the name of the smallest decimal place in that decimal. For example,

- ✔ In the decimal 0.3, the smallest decimal place is the tenths place, so the equivalent fraction is $\frac{3}{10}$.

- ✔ In the decimal 0.29, the smallest decimal place is the hundredths place, so the equivalent fraction is $\frac{29}{100}$.

- ✔ In the decimal 0.817, the smallest decimal place is the hundredths place, so the equivalent fraction is $\frac{817}{1,000}$.

Q. Change the decimal 0.83 to a fraction.

A. $\frac{83}{100}$. Create a "fraction" with 0.83 in the numerator and 1.0 in the denominator:

$$\frac{0.83}{1.0}$$

Move the decimal point in 0.83 two places to the right to turn it into a whole number; move the decimal point in the denominator the same number of places. I do this one decimal place at a time:

$$\frac{0.83}{1.0} = \frac{8.3}{10.0} = \frac{83.0}{100.0}$$

At this point, you can drop the decimal points and trailing zeros in both the numerator and denominator.

Q. Change the decimal 0.0205 to a fraction.

A. $\frac{41}{2,000}$. Create a "fraction" with 0.0205 in the numerator and 1.0 in the denominator:

$$\frac{0.0205}{1.0}$$

Move the decimal point in the 0.0205 four places to the right to turn the numerator into a whole number; move the decimal point in the denominator the same number of places:

$$\frac{0.0205}{1.0}$$
$$= \frac{0.205}{10.0}$$
$$= \frac{02.05}{100.0}$$
$$= \frac{020.5}{1,000.0}$$
$$= \frac{0205.0}{10,000.0}$$

Drop the decimal points, plus any leading or trailing zeros in both the numerator and denominator.

$$= \frac{205}{10,000}$$

Both the numerator and denominator are divisible by 5, so reduce this fraction:

$$= \frac{41}{2,000}$$

23. Change the decimal 0.27 to a fraction.

Solve It

24. Convert the decimal 0.0315 to a fraction.

Solve It

25. Write 45.12 as a mixed number.

Solve It

26. Change 100.001 to a mixed number.

Solve It

Fractions to Decimals

To change any fraction to a decimal, just divide the numerator by the denominator.

Often, you need to find the exact decimal value of a fraction. You can represent every fraction exactly as either a terminating decimal or a repeating decimal:

✔ **Terminating decimal:** A terminating decimal is simply a decimal that has a finite (limited) number of digits. For example, the decimal 0.125 is a terminating decimal with 3 digits. Similarly, the decimal 0.9837596944883383 is a terminating decimal with 16 digits.

✔ **Repeating decimal:** A repeating decimal is a decimal that repeats the same digits forever. For example, the decimal $0.\overline{7}$ is a repeating decimal. The bar over the 7 means that the number 7 is repeated forever: 0.777777777 . . . . Similarly, the decimal $0.34\overline{591}$ is also a repeating decimal. The bar over the 91 means that these two numbers are repeated forever: 0.3459191919191919 . . . .

Whenever the answer to a division problem is a repeating decimal, you'll notice a pattern developing as you divide: When you subtract, you find the same numbers showing up over and over again. When this happens, check the quotient to see whether you can spot the repeating pattern and place a bar over these numbers.

When you're asked to find the exact decimal value of a fraction, feel free to attach trailing zeros to the *dividend* (the number you're dividing) as you go along. Keep dividing until the division either works out evenly (so the quotient is a terminating decimal) or a repeating pattern develops (so it's a repeating decimal).

Q. Convert the fraction $\frac{9}{16}$ to an exact decimal value.

A. **0.5625.** Divide $9 \div 16$:

$$16\overline{)9.000}$$

Because 16 is too big to go into 9, I attached a decimal point and some trailing zeros to the 9. Now you can divide as I show you earlier in this chapter:

$$
\begin{array}{r}
0.5625 \\
16\overline{)9.0000} \\
-80 \\
\hline
100 \\
-96 \\
\hline
40 \\
-32 \\
\hline
80 \\
-80 \\
\hline
0
\end{array}
$$

Q. What is the exact decimal value of the fraction $\frac{5}{6}$?

A. **0.83̄.** Divide $5 \div 6$:

Because 6 is too big to go into 5, I attached a decimal point and some trailing zeros to the 5. Now divide:

$$
\begin{array}{r}
.8333 \\
6\overline{)5.0000} \\
-48 \\
\hline
20 \\
-18 \\
\hline
20 \\
-18 \\
\hline
20 \\
-18 \\
\hline
2
\end{array}
$$

As you can see, a pattern has developed. No matter how many trailing zeros you attach, the quotient will never come out evenly. Instead, the quotient is the repeating decimal .083. The bar over the 3 indicates that the number 3 repeats forever: 0.83333333 . . . .

---

27. Change $\frac{13}{16}$ to an exact decimal value.

Solve It

28. Express $\frac{7}{9}$ exactly as a decimal.

Solve It

Solutions to Getting to the Point with Decimals

The following are the answers to the practice questions presented in this chapter.

1 Expand the following decimals:

a. $2.7 = 2 + \dfrac{7}{10}$

b. $31.4 = 30 + 1 + \dfrac{4}{10}$

c. $86.52 = 80 + 6 + \dfrac{5}{10} + \dfrac{2}{100}$

d. $103.759 = 100 + 3 + \dfrac{7}{10} + \dfrac{5}{100} + \dfrac{9}{1,000}$

e. $1,040.0005 = 1,000 + 40 + \dfrac{5}{10,000}$

f. $16,821.1384 = 10,000 + 6,000 + 800 + 20 + 1 + \dfrac{1}{10} + \dfrac{3}{100} + \dfrac{8}{1,000} + \dfrac{4}{10,000}$

2 Simplify the decimals without removing placeholding zeros:

a. $5.80 = \mathbf{5.8}$

b. $7.030 = \mathbf{7.03}$

c. $90.0400 = \mathbf{90.04}$

d. $9,000.005 = \mathbf{9,000.005}$

e. $0108.0060 = \mathbf{108.006}$

f. $00100.0102000 = \mathbf{100.0102}$

3 Perform decimal multiplication and division:

a. $7.32 \times 10 = \mathbf{73.2}$

b. $9.04 \times 100 = \mathbf{904}$

c. $51.6 \times 100,000 = \mathbf{5,160,000}$

d. $183 \div 100 = \mathbf{1.83}$

e. $2.786 \div 1,000 = \mathbf{0.002786}$

f. $943.812 \div 1,000,000 = \mathbf{0.000943812}$

4 Round each decimal to the number of places indicated:

a. One decimal place: **4.8**

b. Nearest tenth: **52.3**

c. Two decimal places: **191.28**

d. Nearest hundredth: $99.9\underline{95} \rightarrow \mathbf{100.00}$

e. Three decimal places: $0.00\underline{791} \rightarrow \mathbf{0.008}$

f. Nearest thousandth: $909.99\underline{96} \rightarrow \mathbf{910.000}$

5 Convert the decimals to fractions:

a. $0.7 = \dfrac{7}{10}$

b. $0.4 = \dfrac{2}{5}$

c. $0.25 = \dfrac{1}{4}$

d. $0.125 = \dfrac{1}{8}$

e. $0.1 = \dfrac{1}{10}$

f. $0.75 = \dfrac{3}{4}$

6 Change the fractions to decimals:

a. $\dfrac{9}{10} = \mathbf{0.9}$

b. $\dfrac{2}{5} = \mathbf{0.4}$

c. $\dfrac{3}{4} = \mathbf{0.75}$

d. $\dfrac{3}{8} = \mathbf{0.375}$

e. $\dfrac{7}{8} = \mathbf{0.875}$

f. $\dfrac{1}{2} = \mathbf{0.5}$

7 Change the decimals to mixed numbers:

a. $1.6 = \mathbf{1\dfrac{3}{5}}$

b. $3.3 = \mathbf{3\dfrac{3}{10}}$

c. $14.5 = \mathbf{14\dfrac{1}{2}}$

d. $20.75 = \mathbf{20\dfrac{3}{4}}$

e. $100.625 = \mathbf{100\dfrac{5}{8}}$

f. $375.375 = \mathbf{375\dfrac{3}{8}}$

8 Change the mixed numbers to decimals:

a. $1\dfrac{1}{5} = \mathbf{1.2}$

b. $2\dfrac{1}{10} = \mathbf{2.1}$

c. $3\dfrac{1}{2} = \mathbf{3.5}$

d. $5\dfrac{1}{4} = \mathbf{5.25}$

e. $7\dfrac{1}{8} = \mathbf{7.125}$

f. $12\dfrac{5}{8} = \mathbf{12.625}$

9   17.4 + 2.18 = **19.58.** Place the numbers in a column as you would for addition with whole numbers, but with the decimal points lined up. I've filled out the columns with trailing zeros to help show how the columns line up:

$$
\begin{array}{r}
17.40 \\
+2.18 \\
\hline
19.58
\end{array}
$$

Notice that the decimal point in the answer lines up with the others.

10   0.0098 + 10.101 + 0.07 + 33 = **43.1808.** Line up the decimal points and do column addition:

$$
\begin{array}{r}
.00\overset{1}{9}8 \\
10.1010 \\
.0700 \\
+33.0000 \\
\hline
43.1808
\end{array}
$$

11   1,000.001 + 75 + 0.03 + 800.2 = **1,875.231.** Place the decimal numbers in a column, lining up the decimal points:

$$
\begin{array}{r}
1000.001 \\
75.000 \\
0.030 \\
+800.200 \\
\hline
1875.231
\end{array}
$$

12   0.748 – 0.23 = **0.518.** Place the first number on top of the second number, with the decimal points lined up. I've also added a trailing 0 to the second number to fill out the right-hand column and emphasize how the columns line up:

$$
\begin{array}{r}
0.748 \\
-0.230 \\
\hline
0.518
\end{array}
$$

Notice that the decimal point in the answer lines up with the others.

13   674.9 – 5.0001 = **669.8999.** Place the first number on top of the second number, with the decimal points lined up. I've filled out the right-hand column with trailing zeros so I can complete the math:

$$
\begin{array}{r}
6\ \overset{6}{7}\ \overset{1}{4}.\ \overset{8}{9}\ \overset{9}{0}\ \overset{9}{0}\ \overset{1}{0} \\
-5.\ 0\ 0\ 0\ 1 \\
\hline
6\ 6\ 9.\ 8\ 9\ 9\ 9
\end{array}
$$

14   100.009 – 0.68 = **99.329.** Place the first number on top of the second number, with the decimal points lined up:

$$
\begin{array}{r}
\overset{0}{1}\ \overset{1}{0}\ 0.\ \overset{9}{0}\ \overset{9}{0}\ \overset{1}{9} \\
-0.\ 6\ 8\ 0 \\
\hline
9\ 9.\ 3\ 2\ 9
\end{array}
$$

15 $0.635 \times 0.42 = \mathbf{0.2667}$. Place the first number on top of the second number, ignoring the decimal points. Complete the multiplication as you would for whole numbers:

$$0.635 \leftarrow 3 \text{ digits after decimal}$$
$$\underline{\times 0.42} \leftarrow 2 \text{ digits after decimal}$$
$$1270$$
$$\underline{+25400}$$
$$0.26670 \leftarrow 3 + 2 = 5 \text{ digits after decimal}$$

At this point, you're ready to find out where the decimal point goes in the answer. Count the number of decimal places in the two factors, add these two numbers together (3 + 2 = 5), and place the decimal point in the answer so that it has five digits after the decimal point. After you place the decimal point (but not before!), you can drop the trailing zero.

16 $0.675 \times 34.8 = \mathbf{23.49}$. Ignore the decimal points and simply place the first number on top of the second. Complete the multiplication as you would for whole numbers:

$$0.675$$
$$\underline{\times 34.8}$$
$$5400$$
$$27000$$
$$\underline{+202500}$$
$$23.4900$$

Count the number of decimal places in the two factors, add these two numbers together (3 + 1 = 4), and place the decimal point in the answer so that it has four digits after the decimal point. Last, you can drop the trailing zeros.

17 $943 \times 0.0012 = \mathbf{1.1316}$. Complete the multiplication as you would for whole numbers:

$$943 \leftarrow 0 \text{ digit after decimal}$$
$$\underline{\times 0.0012} \leftarrow 4 \text{ digit after decimal}$$
$$1886$$
$$\underline{+9430}$$
$$1.1316 \leftarrow 0 + 4 = 4 \text{ digit after decimal}$$

Zero digits come after the decimal point in the first factor, and you have four after-decimal digits in the second factor, for a total of 4 (0 + 4 = 4); place the decimal point in the answer so that it has four digits after the decimal point.

18 $1.006 \times 0.0807 = \mathbf{0.0811842}$. Complete the multiplication as you would for whole numbers:

$$1.006 \leftarrow 3 \text{ digits after decimal}$$
$$\underline{\times 0.0807} \leftarrow 4 \text{ digits after decimal}$$
$$7042$$
$$\underline{+804800}$$
$$0.0811842 \leftarrow 3 + 4 = 7 \text{ digits after decimal}$$

You have a total of seven digits after the decimal points in the two factors — three in the first factor and four in the second (3 + 4 = 7) — so place the decimal point in the answer so that it has seven digits after the decimal point. Notice that I need to create an extra decimal place in this case by attaching an additional nontrailing 0.

19 9.345 ÷ 0.05 = **186.9.** To start out, write the problem as usual:

$$0.05\overline{)9.345}$$

Turn the divisor (0.05) into a whole number by moving the decimal point two places to the right. At the same time, move the decimal point in the dividend (9.345) two places to the right. Place the decimal point in the quotient directly above where it now appears in the dividend:

$$5.\overline{)934.5}$$

Now you're ready to divide. Be careful to line up the quotient properly so that the decimal point falls into place.

$$
\begin{array}{r}
186.9 \\
5.\overline{)934.5} \\
-5 \\
\hline
43 \\
-40 \\
\hline
34 \\
-30 \\
\hline
45 \\
-45 \\
\hline
0
\end{array}
$$

20 3.15 ÷ 0.021 = **150.** Write the problem as usual:

$$0.021\overline{)3.15}$$

You need to move the decimal point in the divisor (0.021) three places to the right, so attach an additional trailing zero to the dividend (3.15) to extend it to three decimal places:

$$0.021\overline{)3.150}$$

Now you can move both decimal points three places to the right. Place the decimal point in the quotient above the decimal point in the dividend:

$$21.\overline{)3150.}$$

Divide, being careful to line up the quotient properly:

$$
\begin{array}{r}
150. \\
21\overline{)3150.} \\
-21 \\
\hline
105 \\
-105 \\
\hline
0
\end{array}
$$

Remember to insert a placeholding zero in the quotient so that the decimal point ends up in the correct place.

21 $6.7 \div 10.1 =$ **0.7.** To start out, write the problem as usual:

$$10.1 \overline{)6.7}$$

Turn the divisor (10.1) into a whole number by moving the decimal point one place to the right. At the same time, move the decimal point in the dividend (6.7) one place to the right:

$$101. \overline{)67.}$$

The problem asks you to round the quotient to one decimal place, so fill out the dividend with trailing zeros to two decimal places:

$$101. \overline{)67.00}$$

Now you're ready to divide:

$$
\begin{array}{r}
0.66 \\
101. \overline{)67.00} \\
-606 \\
\hline
640 \\
-606 \\
\hline
34
\end{array}
$$

Round the quotient to one decimal place:

$$0.6\underline{6} \rightarrow 0.7$$

22 $9.13 \div 4.25 =$ **2.15.** First, write the problem as usual:

$$4.25 \overline{)9.13}$$

Turn the divisor (4.25) into a whole number by moving the decimal point two places to the right. At the same time, move the decimal point in the dividend (9.13) two places to the right:

$$425 \overline{)913.}$$

The problem asks you to round the quotient to the nearest hundredth, so fill out the dividend with trailing zeros to three decimal places:

$$425. \overline{)913.000}$$

Now divide, carefully lining up the quotient:

$$
\begin{array}{r}
2.148 \\
425. \overline{)913.000} \\
-850 \\
\hline
630 \\
-425 \\
\hline
2050 \\
-1700 \\
\hline
3500 \\
-3400 \\
\hline
100
\end{array}
$$

Round the quotient to the nearest hundredth:

$2.1\underline{48} \rightarrow 2.15$

23 $0.27 = \dfrac{27}{100}$. Create a "fraction" with 0.27 in the numerator and 1.0 in the denominator. Then move the decimal points to the right until both the numerator and denominator are whole numbers:

$$\frac{0.27}{1.0} = \frac{2.7}{10.0} = \frac{27.0}{100.0}$$

At this point, you can drop the decimal points and trailing zeros.

24 $0.0315 = \dfrac{63}{2,000}$. Create a "fraction" with 0.0315 in the numerator and 1.0 in the denominator. Then move the decimal points in both the numerator and denominator to the right one place at a time. Continue until both the numerator and denominator are whole numbers:

$$\frac{0.0315}{1.0} = \frac{0.315}{10.0} = \frac{3.15}{100.0} = \frac{31.5}{1,000.0} = \frac{315.0}{10,000.0}$$

Drop the decimal points and trailing zeros. The numerator and denominator are both divisible by 5, so reduce the fraction:

$$\frac{315}{10,000} = \frac{63}{2,000}$$

25 $45.12 = 45\dfrac{3}{25}$. Before you begin, separate out the whole-number portion of the decimal (45). Create a "fraction" with 0.12 in the numerator and 1.0 in the denominator. Move the decimal points in both the numerator and denominator to the right until both are whole numbers:

$$\frac{0.12}{1.0} = \frac{1.2}{10.0} = \frac{12.0}{100.0}$$

Drop the decimal points and trailing zeros. As long as the numerator and denominator are both divisible by 2 (that is, even numbers), you can reduce this fraction:

$$\frac{12}{100} = \frac{6}{50} = \frac{3}{25}$$

To finish up, reattach the whole-number portion that you separated at the beginning.

26 $100.001 = 100\dfrac{1}{1,000}$. Separate out the whole-number portion of the decimal (100) and create a "fraction" with 0.001 in the numerator and 1.0 in the denominator. Move the decimal points in both the numerator and denominator to the right one place at a time until both are whole numbers:

$$\frac{0.001}{1.0} = \frac{0.01}{10.0} = \frac{0.1}{100.0} = \frac{1.0}{1,000.0}$$

Drop the decimal points and trailing zeros and reattach the whole-number portion of the number you started with:

$$100\frac{1}{1,000}$$

27 $\frac{13}{16} = $ **0.8125.** Divide $13 \div 16$, attaching plenty of trailing zeros to the 13:

$$
\begin{array}{r}
0.8125 \\
16\overline{)13.00000} \\
\underline{-128} \\
20 \\
\underline{-16} \\
40 \\
\underline{-32} \\
80 \\
\underline{-80} \\
0
\end{array}
$$

This division eventually ends, so the quotient is a terminating decimal.

28 $\frac{7}{9} = $ **0.$\overline{7}$.** Divide $7 \div 9$, attaching plenty of trailing zeros to the 7:

$$
\begin{array}{r}
0.77 \\
9\overline{)7.000} \\
\underline{-63} \\
70 \\
\underline{-63} \\
70
\end{array}
$$

A pattern has developed in the subtraction: $70 - 63 = 7$, so when you bring down the next 0, you'll get 70 again. Therefore, the quotient is a repeating decimal.

Chapter 9

Playing the Percentages

- -

In This Chapter

▶ Converting between percents and decimals

▶ Switching between percents and fractions

▶ Solving all three types of percent problems

- -

*L*ike fractions and decimals (which I discuss in Chapters 6, 7, and 8), percents are a way of describing parts of the whole. The word *percent* literally means "for 100," but in practice, it means "out of 100." So, when I say that 50 percent of my shirts are blue, I mean that 50 out of 100 shirts — that is, half of them — are blue. Of course, you don't really have to own as many shirts as I claim to in order to make this statement. If you own 8 shirts and 50 percent of them are blue, then you own 4 blue shirts.

In this chapter, I show you how to convert percents to and from decimals and fractions. In the last section, I show you how to translate percent problems into equations to solve the three main types of percent problems. Become familiar with percents, and you'll be able to figure out discounts, sales tax, tips for servers, and my favorite: interest on money in the bank.

Converting Percents to Decimals

Percents and decimals are very similar forms, so everything you know about decimals (see Chapter 8) carries over when you're working with percents. All you need to do is convert your percent to a decimal, and you're good to go.

To change a whole-number percent to a decimal, simply replace the percent sign with a decimal point and then move this decimal point two places to the left; after this, you can drop any trailing zeros. Here are a few common conversions between percents and decimals:

100% = 1	75% = 0.75	50% = 0.5
25% = 0.25	20% = 0.2	10% = 0.1

Sometimes a percent already has a decimal point. In this case, just drop the percent sign and move the decimal point two places to the left. For instance, 12.5% = 0.125.

Q. Change 80% to a decimal.

A. **0.8.** Replace the percent sign with a decimal point — changing 80% to 80. — and then move the decimal point two places to the left:

$$80\% = 0.80$$

At the end, you can drop the trailing zero to get 0.8.

Q. Change 37.5% to a decimal.

A. **0.375.** Drop the percent sign and move the decimal point two places to the left:

$$37.5\% = 0.375$$

1. Change 90% to a decimal.

Solve It

2. A common interest rate on an investment such as a bond is 4%. Convert 4% to a decimal.

Solve It

3. Find the decimal equivalent of 99.44%.

Solve It

4. What is 243.1% expressed as a decimal?

Solve It

Changing Decimals to Percents

Calculating with percents is easiest when you convert to decimals first. When you're done calculating, however, you often need to change your answer from a decimal back to a percent. This is especially true when you're working with interest rates, taxes, or the likelihood of a big snowfall the night before a big test. All these numbers are most commonly expressed as percents.

To change a decimal to a percent, move the decimal point two places to the right and attach a percent sign. If the result is a whole number, you can drop the decimal point.

Q. Change 0.6 to a percent.

A. **60%.** Move the decimal point two places to the right and attach a percent sign:

$$0.6 = 60\%$$

---

5. Convert 0.57 to a percent.

Solve It

---

6. What is 0.3 expressed as a percent?

Solve It

---

7. Change 0.015 to a percent.

Solve It

---

8. Express 2.222 as a percent.

Solve It

Switching from Percents to Fractions

Some percents are easy to convert to fractions. Here are a few quick conversions that are worth knowing:

$$1\% = \frac{1}{100} \qquad 5\% = \frac{1}{20} \qquad 10\% = \frac{1}{10} \qquad 20\% = \frac{1}{5}$$

$$25\% = \frac{1}{4} \qquad 50\% = \frac{1}{2} \qquad 75\% = \frac{3}{4} \qquad 100\% = 1$$

Beyond these simple conversions, changing a percent to a fraction isn't a skill you're likely to use much outside of a math class. Decimals are much easier to work with.

However, teachers often test you on this skill to make sure you understand the ins and outs of percents, so here's the scoop on converting percents to fractions: To change a percent to a fraction, use the percent without the percent sign as the *numerator* (top number) of the fraction and use 100 as the *denominator* (bottom number). When necessary, reduce this fraction to lowest terms or change it to a mixed number. (For a refresher on reducing fractions, see Chapter 6.)

Q. Change 35% to a fraction.

A. $\frac{7}{20}$. Place 35 in the numerator and 100 in the denominator:

$$35\% = \frac{35}{100}$$

You can reduce this fraction because the numerator and denominator are both divisible by 5:

$$\frac{7}{20}$$

9. Change 19% to a fraction.

Solve It

10. A common interest rate on credit cards and other types of loans is 8%. What is 8% expressed as a fraction?

Solve It

11. Switch 123% to a fraction.

Solve It

12. Convert 375% to a fraction.

Solve It

Converting Fractions to Percents

Knowing how to make a few simple conversions from fractions to percents is a useful real-world skill. Here are some of the most common conversions:

$$\frac{1}{100} = 1\% \qquad \frac{1}{20} = 5\% \qquad \frac{1}{10} = 10\% \qquad \frac{1}{5} = 20\%$$

$$\frac{1}{4} = 25\% \qquad \frac{1}{2} = 50\% \qquad \frac{3}{4} = 75\% \qquad 1 = 100\%$$

Beyond these, you're not all that likely to need to convert a fraction to a percent outside of a math class. But then, passing your math class is important, so in this section I show you how to make this type of conversion.

Converting a fraction to a percent is a two-step process:

1. **Convert the fraction to a decimal, as I show you in Chapter 8.**

   In some problems, the result of this step may be a repeating decimal. This is fine — in this case, the percent will also contain a repeating decimal.

2. **Convert this decimal to a percent.**

   Move the decimal point two places to the right and add a percent sign.

Q. Change the fraction $\frac{1}{9}$ to a percent.

A. **11.$\bar{1}$%**. First, change $\frac{1}{9}$ to a decimal:

$$
\begin{array}{r}
0.111 \\
9\overline{)1.000} \\
\underline{-9} \\
10 \\
\underline{-9} \\
10 \\
\underline{-9} \\
1
\end{array}
$$

The result is the repeating decimal $0.\bar{1}$. Now change this repeating decimal to a percent:

$$0.\bar{1} = 11.\bar{1}\%$$

---

13. Express $\frac{2}{5}$ as a percent.

Solve It

14. Change $\frac{3}{20}$ to a percent.

Solve It

---

15. Convert $\frac{7}{8}$ to a percent.

Solve It

16. Change $\frac{2}{11}$ to a percent.

Solve It

Solving a Variety of Percent Problems Using Word Equations

In this section, I show you how to recognize the three main types of percent problems. Then I show you how to solve them all using word equations.

Percent problems give you two pieces of information and ask you to find the third piece. Here are the three pieces — and the kinds of questions that ask for each piece:

- **The percent:** The problem may give you the starting and ending numbers and ask you to find the percent. Here are some ways this problem can be asked:

    ?% of 4 is 1

    What percent of 4 is 1?

    1 is what percent of 4?

    The answer is 25%, because 25% × 4 = 1.

- **The starting number:** The problem may give you the percent and the ending number and ask you to find the starting number:

    10% of ? is 40

    10% of what number is 40?

    40 is 10% of what number?

    This time, the answer is 400, because 10% × 400 = 40.

- **The ending number:** The most common type of percent problem gives you a percentage and a starting number and asks you to figure out the ending number:

    50% of 6 is ?

    50% of 6 equals what number?

    Can you find 50% of 6?

    No matter how I phrase it, notice that the problem always includes *50% of 6*. The answer is 3, because 50% × 6 = 3.

Each type of percent problem gives you *two* pieces of information and asks you to find the third. Place the information into an equation using the following translations from words into symbols:

What (number) → n

is → =

percent → × 0.01

of → ×

Q. Place the statement 25% of 12 is 3 into an equation.

A. **25 × 0.01 × 12 = 3.**

This is a direct translation, as follows:

25	%	of	12	is	3
25	× 0.01	×	12	=	3

Q. What is 18% of 90?

A. **16.2.** Translate the problem into an equation:

What	is	18	%	of	90
n	=	18	× 0.01	×	90

Solve this equation:

$n = 18 × 0.01 × 90 = 16.2$

---

17. Place the statement *20% of 350 is 70* into an equation. Check your work by simplifying the equation.

Solve It

18. What percent of 150 is 25.5?

Solve It

---

19. What is 79% of 11?

Solve It

20. 30% of what number is 10?

Solve It

Solutions to Playing the Percentages

The following are the answers to the practice questions presented in this chapter.

1 **0.9.** Replace the percent sign with a decimal point and then move this decimal point two places to the left:

90% = 0.90

At the end, drop the trailing zero to get 0.9.

2 **0.04.** Replace the percent sign with a decimal point and then move this decimal point two places to the left:

4% = 0.04

3 **0.9944.** Drop the percent sign and move the decimal point two places to the left:

99.44% = 0.9944

4 **2.431.** Drop the percent sign and move the decimal point two places to the left:

243.1% = 2.431

5 **57%.** Move the decimal point two places to the right and attach a percent sign:

0.57 = 057%

At the end, drop the leading zero to get 57%.

6 **30%.** Move the decimal point two places to the right and attach a percent sign:

0.3 = 030%

At the end, drop the leading zero to get 30%.

7 **1.5%.** Move the decimal point two places to the right and attach a percent sign:

0.015 = 01.5%

At the end, drop the leading zero to get 1.5%.

8 **222.2%.** Move the decimal point two places to the right and attach a percent sign:

2.222 = 222.2%

9 $\frac{19}{100}$. Place 19 in the numerator and 100 in the denominator.

10 $\frac{2}{25}$. Place 8 in the numerator and 100 in the denominator:

$$\frac{8}{100}$$

You can reduce this fraction by 2 twice:

$$= \frac{4}{50} = \frac{2}{25}$$

11 $1\frac{23}{100}$. Place 123 in the numerator and 100 in the denominator:

$$\frac{123}{100}$$

You can change this improper fraction to a mixed number:

$$= 1\frac{23}{100}$$

12 $3\frac{3}{4}$. Place 375 in the numerator and 100 in the denominator:

$$\frac{375}{100}$$

Change the improper fraction to a mixed number:

$$= 3\frac{75}{100}$$

Reduce the fractional part of this mixed number, first by 5 and then by another 5:

$$= 3\frac{15}{20} = 3\frac{3}{4}$$

13 **40%.** First, change $\frac{2}{5}$ to a decimal:

$$2.0 \div 5 = 0.4$$

Now change 0.4 to a percent by moving the decimal point two places to the right and adding a percent sign:

$$0.4 = 40\%$$

14 **15%.** First, change $\frac{3}{20}$ to a decimal:

$$3.00 \div 20 = 0.15$$

Then change 0.15 to a percent:

$$0.15 = 15\%$$

15 **87.5%.** First, change $\frac{7}{8}$ to a decimal:

$$7.000 \div 8 = 0.875$$

Now change 0.875 to a percent:

$$0.875 = 87.5\%$$

16   **18.$\overline{18}$%.** First, change $\frac{2}{11}$ to a decimal:

$$
\begin{array}{r}
0.1818 \\
11\overline{)2.0000} \\
-11 \\
\hline
90 \\
-88 \\
\hline
20 \\
-11 \\
\hline
90 \\
-88 \\
\hline
2
\end{array}
$$

The result is the repeating decimal $0.\overline{18}$. Now change this repeating decimal to a percent:

$$0.\overline{18} = 18.\overline{18}\%$$

17     $20 \times 0.01 \times 350 =$ **70.**

Turn the problem into an equation as follows:

20	percent	of	350	is	70
20	× 0.01	×	350	=	70

Check this equation:

$$
\begin{aligned}
20 \times 0.01 \times 350 &= 70 \\
0.2 \times 350 &= 70 \\
70 &= 70
\end{aligned}
$$

18   **17%.**

Turn the problem into an equation:

What	percent	of	150	is	25.5
n	× 0.01	×	150	=	25.5

Solve the equation for n:

$$
\begin{aligned}
n \times 0.01 \times 150 &= 25.5 \\
1.5n &= 25.5 \\
\frac{1.5n}{1.5} &= \frac{25.5}{1.5} \\
n &= 17
\end{aligned}
$$

19 **8.69.**

Turn the problem into an equation:

What	is	79	%	of	11
n	=	79	$\times 0.01$	$\times$	11

To find the answer, solve this equation for n:

$$n = 79 \times 0.01 \times 11 = 8.69$$

20 **$33.\overline{3}$.**

Turn the problem into an equation:

30	%	of	what number	is	10
30	$\times 0.01$	$\times$	n	=	10

Solve for n:

$$30 \times 0.01 \times n = 10$$
$$0.3n = 10$$
$$\frac{0.3n}{0.3} = \frac{10}{0.3}$$
$$n = 33.\overline{3}$$

The answer is the repeating decimal $33.\overline{3}$.

Part III
A Giant Step Forward: Intermediate Topics

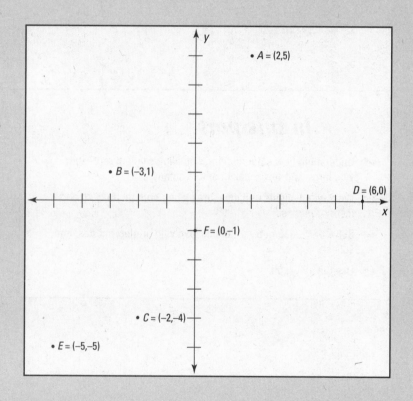

Make sense of weird exponents in an article at www.dummies.com/extras/ basicmathprealgebrawb.

In this part. . .

✔ Understand how scientific notation allows you to represent very large and very small numbers simply.

✔ Work with weights and measures using both the English and metric systems.

✔ Solve basic geometry problems involving angles, shapes, and solids.

✔ Use the *xy*-graph.

Chapter 10

Seeking a Higher Power through Scientific Notation

Powers of ten — the number 10 multiplied by itself any number of times — form the basis of the Hindu-Arabic number system (the decimal number system) that you're familiar with. In Chapter 1, you discover how this system uses zeros as placeholders, giving you the ones place, the tens place, the ten-trillions place, and the like. This system works well for relatively small numbers, but as numbers grow, using zeros becomes awkward. For example, ten quintillion is represented as the number 10,000,000,000,000,000,000.

Similarly, in Chapter 8, you find zeros also work as placeholders in decimals. In this case, the system works great for decimals that aren't overly precise, but it becomes awkward when you need a high level of precision. For example, *two trillionths* is represented as the decimal 0.000000000002.

And really, people are busy, so who has time to write out all those place-holding zeros when they could be birdwatching, baking a pie, developing a shark-robot security system, or something else that's more fun? Well, now you can skip some of those zeros and devote your time to more important pursuits. In this chapter, you discover *scientific notation* as a handy alternative way of writing very large numbers and very small decimals. Not surprisingly, scientific notation is most commonly used in the sciences, where big numbers and precise decimals show up all the time.

On the Count of Zero: Understanding Powers of Ten

As you discover in Chapter 2, raising a number to a *power* multiplies the number in the base (the bottom number) by itself as many times as indicated by the exponent (the top number). For example, $2^3 = 2 \times 2 \times 2 = 8$.

Powers often take a long time to calculate because the numbers grow so quickly. For example, 7^6 may look small, but it equals 117,649. But the easiest powers to calculate are powers with a base of 10 — called, naturally, the *powers of ten*. You can write every power of ten in two ways:

✔ **Standard notation:** As a number, such as 100,000,000

✔ **Exponential notation:** As the number 10 raised to a power, such as 10^8

Powers of ten are easy to spot, because in standard notation, every power of ten is simply the digit 1 followed by all 0s. To raise 10 to the power of any number, just write a 1 with that number of 0s after it. For example,

$10^0 = 1$	1 with no 0s
$10^1 = 10$	1 with one 0
$10^2 = 100$	1 with two 0s
$10^3 = 1,000$	1 with three 0s

To switch from standard to exponential notation, you simply count the zeros and use that as the exponent on the number 10.

You can also raise 10 to the power of a negative number. The result of this operation is always a decimal, with the 0s coming before the 1. For example,

$10^{-1} = 0.1$	1 with one 0
$10^{-2} = 0.01$	1 with two 0s
$10^{-3} = 0.001$	1 with three 0s
$10^{-4} = 0.0001$	1 with four 0s

When expressing a negative power of ten in standard form, always count the leading 0 — that is, the 0 to the left of the decimal point. For example, $10^{-3} = 0.001$ has three 0s, counting the leading 0.

Q. Write 10^6 in standard notation.

A. **1,000,000.** The exponent is 6, so the standard notation is a 1 with six 0s after it.

Q. Write 100,000 in exponential notation.

A. 10^5. The number 100,000 has five 0s, so the exponential notation has 5 in the exponent.

Q. Write 10^{-5} in standard notation.

A. **0.00001.** The exponent is –5, so the standard notation is a decimal with five 0s (including the leading 0) followed by a 1.

Q. Write 0.0000001 in exponential notation.

A. 10^{-7}. The decimal has seven 0s (including the leading 0), so the exponential notation has –7 in the exponent.

1. Write each of the following powers of ten in standard notation:

    a. 10^4

    b. 10^7

    c. 10^{14}

    d. 10^{22}

Solve It

2. Write each of the following powers of ten in exponential notation:

    a. 1,000,000,000

    b. 1,000,000,000,000

    c. 10,000,000,000,000,000

    d. 100,000,000,000,000,000,000,000,000,000,000

Solve It

3. Write each of the following powers of ten in standard notation:

    a. 10^{-1}

    b. 10^{-5}

    c. 10^{-11}

    d. 10^{-16}

Solve It

4. Write each of the following powers of ten in exponential notation:

    a. 0.01

    b. 0.000001

    c. 0.000000000001

    d. 0.000000000000000001

Solve It

Exponential Arithmetic: Multiplying and Dividing Powers of Ten

Multiplying and dividing powers of ten in exponential notation is a snap because you don't have to do any multiplying or dividing at all — it's nothing more than simple addition and subtraction:

- **Multiplication:** To multiply two powers of ten in exponential notation, find the sum of the numbers' exponents; then write a power of ten using that sum as the exponent.

- **Division:** To divide one power of ten by another, subtract the second exponent from the first; then write a power of ten using this resulting sum as the exponent.

This rule works equally well when one or both exponents are negative — just use the rules for adding negative numbers, which I discuss in Chapter 3.

Q. Multiply 10^7 by 10^4.

A. 10^{11}. Add the exponents $7 + 4 = 11$, and use this as the exponent of your answer:

$$10^7 \times 10^4 = 10^{7+4} = 10^{11}$$

Q. Find $10^9 \div 10^6$.

A. 10^3. For division, you subtract. Subtract the exponents $9 - 6 = 3$ and use this as the exponent of your answer:

$$10^9 \div 10^6 = 10^{(9-6)} = 10^3$$

5. Multiply each of the following powers of ten:

a. $10^9 \times 10^2$

b. $10^5 \times 10^5$

c. $10^{13} \times 10^{-16}$

d. $10^{100} \times 10^{21}$

e. $10^{-15} \times 10^0$

f. $10^{-10} \times 10^{-10}$

Solve It

6. Divide each of the following powers of ten:

a. $10^6 \div 10^4$

b. $10^{12} \div 10^1$

c. $10^{-7} \div 10^{-7}$

d. $10^{18} \div 10^0$

e. $10^{100} \div 10^{-19}$

f. $10^{-50} \div 10^{50}$

Solve It

Representing Numbers in Scientific Notation

Numbers with a lot of zeros are awkward to work with, and making mistakes with them is easy. *Scientific notation* is a clearer, alternative way of representing large and small numbers. Every number can be represented in scientific notation as the *product* of two numbers (two numbers multiplied together):

> ✔ A decimal greater than or equal to 1 and less than 10

> ✔ A power of ten written in exponential form

Use the following steps to write any number in scientific notation:

1. **Write the number as a decimal (if it isn't one already) by attaching a decimal point and one trailing zero.**

2. **Move the decimal point just enough places to change this decimal to a new decimal that's greater than or equal to 1 and less than 10. (Be sure to count the number of places moved.)**

   One nonzero digit should come to the left of the decimal point.

3. **Multiply the new decimal by 10 raised to the power that equals the number of places you moved the decimal point in Step 2.**

4. **If you moved the decimal point to the left in Step 2, the exponent is positive. If you moved it to the right (your original number was less than 1), put a minus sign on the exponent.**

Q. Change the number 70,000 to scientific notation.

A. 7.0×10^4. First, write the number as a decimal:

70,000.0

Move the decimal point just enough places to change this decimal to a new decimal that's between 1 and 10. In this case, move the decimal four places to the left. You can drop all but one trailing zero:

7.0

You moved the decimal point four places, so multiply the new number by 10^4:

7.0×10^4

Because you moved the decimal point to the left (you started with a big number), the exponent is a positive number, so you're done.

Q. Change the decimal 0.000000439 to scientific notation.

A. 4.39×10^{-7}. You're starting with a decimal, so Step 1 — writing the number as a decimal — is already taken care of:

0.000000439

To change 0.000000439 to a decimal that's between 1 and 10, move the decimal point seven places to the right and drop the leading zeros:

4.39

Because you moved seven places to the right, multiply the new number by 10^{-7}:

4.39×10^{-7}

7. Change 2,591 to scientific notation.

Solve It

8. Write the decimal 0.087 in scientific notation.

Solve It

9. Write 1.00000783 in scientific notation.

Solve It

10. Convert 20,002.00002 to scientific notation.

Solve It

Multiplying and Dividing with Scientific Notation

Because scientific notation keeps track of place-holding zeros for you, multiplying and dividing by scientific notation is really a lot easier than working with big numbers and tiny decimals that have tons of zeros.

To multiply two numbers in scientific notation, follow these steps:

1. **Multiply the two decimal parts to find the decimal part of the answer.**

   See Chapter 8 for info on multiplying decimals.

2. **Add the exponents on the 10s to find the power of ten in the answer.**

   You're simply multiplying the powers of ten, as I show you earlier in "Exponential Arithmetic: Multiplying and Dividing Powers of Ten."

3. **If the decimal part of the result is 10 or greater, adjust the result by moving the decimal point one place to the left and adding 1 to the exponent.**

Here's how to divide two numbers in scientific notation:

1. **Divide the decimal part of the first number by the decimal part of the second number to find the decimal part of the answer.**

2. **To find the power of ten in the answer, subtract the exponent on the second power of ten from the exponent on the first.**

   You're really just dividing the first power of ten by the second.

3. **If the decimal part of the result is less than 1, adjust the result by moving the decimal point one place to the right and subtracting 1 from the exponent.**

Q. Multiply 2.0×10^3 by 4.1×10^4.

A. 8.2×10^7. Multiply the two decimal parts:

$2.0 \times 4.1 = 8.2$

Then multiply the powers of ten by adding the exponents:

$10^3 \times 10^4 = 10^{3+4} = 10^7$

In this case, no adjustment is necessary because the resulting decimal part is less than 10.

Q. Divide 3.4×10^4 by 2.0×10^9.

A. 1.7×10^{-5}. Divide the first decimal part by the second:

$3.4 \div 2.0 = 1.7$

Then divide the first power of ten by the second by subtracting the exponents:

$10^4 \div 10^9 = 10^{4-9} = 10^{-5}$

In this case, no adjustment is necessary because the resulting decimal part isn't less than 1.

11. Multiply 1.5×10^7 by 6.0×10^5.

Solve It

12. Divide 6.6×10^8 by 1.1×10^3.

Solve It

Answers to Problems in Seeking a Higher Power through Scientific Notation

The following are the answers to the practice questions presented in this chapter.

1 In each case, write the digit 1 followed by the number of 0s indicated by the exponent:

a. 10^4 = **10,000**

b. 10^7 = **10,000,000**

c. 10^{14} = **100,000,000,000,000**

d. 10^{22} = **10,000,000,000,000,000,000,000**

2 In each case, count the number of 0s; then write a power of ten with this number as the exponent.

a. 1,000,000,000 = **10^9**

b. 1,000,000,000,000 = **10^{12}**

c. 10,000,000,000,000,000 = **10^{16}**

d. 100,000,000,000,000,000,000,000,000,000,000 = **10^{32}**

3 Write a decimal beginning with all 0s and ending in 1. The exponent indicates the number of 0s in this decimal (including the leading 0):

a. 10^{-1} = **0.1**

b. 10^{-5} = **0.00001**

c. 10^{-11} = **0.00000000001**

d. 10^{-16} = **0.0000000000000001**

4 In each case, count the number of 0s (including the leading 0); then write a power of ten using this number *negated* (with a minus sign) as the exponent:

a. 0.01 = **10^{-2}**

b. 0.000001 = **10^{-6}**

c. 0.000000000001 = **10^{-12}**

d. 0.000000000000000001 = **10^{-18}**

5 Add the exponents and use this sum as the exponent of the answer.

a. $10^9 \times 10^2 = 10^{9+2} = \mathbf{10^{11}}$

b. $10^5 \times 10^5 = 10^{5+5} = \mathbf{10^{10}}$

c. $10^{13} \times 10^{-16} = 10^{13+-16} = \mathbf{10^{-3}}$

d. $10^{100} \times 10^{21} = 10^{100+21} = \mathbf{10^{121}}$

e. $10^{-15} \times 10^0 = 10^{-15+0} = \mathbf{10^{-15}}$

f. $10^{-10} \times 10^{-10} = 10^{-10+-10} = \mathbf{10^{-20}}$

6   In each case, subtract the first exponent minus the second and use this result as the exponent of the answer.

     a. $10^6 \div 10^4 = 10^{6-4} = \mathbf{10^2}$

     b. $10^{12} \div 10^1 = 10^{12-1} = \mathbf{10^{11}}$

     c. $10^{-7} \div 10^{-7} = 10^{-7-(-7)} = 10^{-7+7} = \mathbf{10^0}$

     d. $10^{18} \div 10^0 = 10^{18-0} = \mathbf{10^{18}}$

     e. $10^{100} \div 10^{-19} = 10^{100-(-19)} = 10^{100+19} = \mathbf{10^{119}}$

     f. $10^{-50} \div 10^{50} = 10^{-50-50} = \mathbf{10^{-100}}$

7    $2{,}591 = \mathbf{2.591 \times 10^3}$. Write 2,591 as a decimal:

     2,591.0

To change 2,591.0 to a decimal between 1 and 10, move the decimal point three places to the left and drop the trailing zero:

     2.591

Because you moved the decimal point three places, multiply the new decimal by 10^3:

     2.591×10^3

You moved the decimal point to the left, so the exponent stays positive. The answer is 2.591×10^3.

8    $0.087 = \mathbf{8.7 \times 10^{-2}}$. To change 0.087 to a decimal between 1 and 10, move the decimal point two places to the right and drop the leading zero:

     8.7

Because you moved the decimal point two places to the right, multiply the new decimal by 10^{-2}:

     8.7×10^{-2}

9    $1.00000783 = \mathbf{1.00000783}$. The decimal 1.00000783 is already between 1 and 10, so no change is needed.

10    $20{,}002.00002 = \mathbf{2.000200002 \times 10^4}$. The number 20,002.00002 is already a decimal. To change it to a decimal between 1 and 10, move the decimal point four places to the left:

     2.000200002

Because you moved the decimal point four places, multiply the new decimal by 10^4:

     2.000200002×10^4

You moved the decimal point to the left, so the answer is 2.000200002×10^4.

11 $(1.5 \times 10^7)(6.0 \times 10^5) = \mathbf{9.0 \times 10^{12}}$. Multiply the two decimal parts:

$$1.5 \times 6.0 = 9.0$$

Multiply the two powers of ten:

$$10^7 \times 10^5 = 10^{7+5} = 10^{12}$$

In this case, no adjustment is necessary because the decimal is less than 10.

12 $(6.6 \times 10^8) \div (1.1 \times 10^3) = \mathbf{6.0 \times 10^5}$. Divide the first decimal part by the second:

$$6.6 \div 1.1 = 6.0$$

Divide the first power of ten by the second:

$$10^8 \div 10^3 = 10^{8-3} = 10^5$$

In this case, no adjustment is necessary because the decimal is greater than 1.

Chapter 11

Weighty Questions on Weights and Measures

*U*nits connect numbers to the real world. For example, the number 2 is just an idea until you attach a unit: 2 apples, 2 children, 2 houses, 2 giant squid, and so forth. Apples, children, houses, and giant squid are easy to do math with because they're all discrete — that is, they're separate and easy to count one by one. For example, if you're working with a basket of apples, applying the Big Four operations is quite straightforward. You can add a few apples to the basket, divide them into separate piles, or perform any other operation that you like.

However, lots of things aren't discrete but rather *continuous* — that is, they're difficult to separate and count one by one. To measure the length of a road, the amount of water in a bucket, the weight of a child, the amount of time a job takes to do, or the temperature of Mount Erebus (and other Antarctic volcanoes), you need *units of measurement*.

The two most common systems of measurement are the English system (used in the United States) and the metric system (used throughout the rest of the world). In this chapter, I familiarize you with both systems. Then I show you how to make conversions between both systems.

The Basics of the English System

The English system of measurement is most commonly used in the United States. If you were raised in the States, you're probably familiar with most of these units. Table 11-1 shows you the most common units and some equations so you can do simple conversions from one unit to another.

Table 11-1	Commonly Used English Units of Measurement	
Measure of	*English Units*	*Conversion Equations*
Distance (length)	Inches (in.)	
	Feet (ft.)	12 inches = 1 foot
	Yards (yd.)	3 feet = 1 yard
	Miles (mi.)	5,280 feet = 1 mile
Fluid volume (capacity)	Fluid ounces (fl. oz.)	
	Cups (c.)	8 fluid ounces = 1 cup
	Pints (pt.)	2 cups = 1 pint
	Quarts (qt.)	2 pints = 1 quart
	Gallons (gal.)	4 quarts = 1 gallon
Weight	Ounces (oz.)	
	Pounds (lb.)	16 ounces = 1 pound
	Tons	2,000 pounds = 1 ton
Time	Seconds	
	Minutes	60 seconds = 1 minute
	Hours	60 minutes = 1 hour
	Days	24 hours = 1 day
	Weeks	7 days = 1 week
	Years	365 days = 1 year
Temperature	Degrees Fahrenheit (°F)	
Speed (rate)	Miles per hour (mph)	

To use this chart, remember the following rules:

✔ When converting from a large unit to a smaller unit, always multiply. For example, 2 pints are in 1 quart, so to convert 10 quarts to pints, multiply by 2:

10 quarts × 2 = 20 pints

✔ When converting from a small unit to a larger unit, always divide. For example, 3 feet are in 1 yard, so to convert 12 feet to yards, divide by 3:

12 feet ÷ 3 = 4 yards

When converting from large units to very small ones (for example, from tons to ounces), you may need to multiply more than once. Similarly, when converting from small units to much larger ones (for example, from minutes to days), you may need to divide more than once.

After doing a conversion, step back and apply a *reasonability test* to your answer — that is, think about whether your answer makes sense. For example, when you convert feet to inches, the number you end up with should be a lot bigger than the number you started with because there are lots of inches in a foot.

Q. How many minutes are in a day?

A. **1,440 minutes.** You're going from a larger unit (a day) to a smaller unit (minutes); 60 minutes are in an hour and 24 hours are in a day, so you simply multiply:

60 × 24 = 1,440 minutes

Therefore, 1,440 minutes are in a day.

Q. 5 pints = _____ fluid ounces.

A. **80 fluid ounces.** 8 fluid ounces are in a cup and 2 cups are in a pint, so

8 × 2 = 16 fl. oz.

16 fluid ounces are in a pint, so

5 pints = 5 × 16 fluid ounces = 80 fl. oz.

Q. If you have 32 fluid ounces, how many pints do you have?

A. **2 pints.** 8 fluid ounces are in a cup, so divide as follows:

32 fluid ounces ÷ 8 = 4 cups

And 2 cups are in a pint, so divide again:

4 cups ÷ 2 = 2 pints

Therefore, 32 fluid ounces equals 2 pints.

Q. 504 hours = _____ weeks.

A. **3 weeks.** 24 hours are in a day, so divide as follows:

504 hours ÷ 24 = 21 days

And 7 days are in a week, so divide again:

21 days ÷ 7 = 3 weeks

Therefore, 504 hours equals 3 weeks.

---

1. Answer each of the following questions:

a. How many inches are in a yard?

b. How many hours are in a week?

c. How many ounces are in a ton?

d. How many cups are in a gallon?

Solve It

2. Calculate each of the following:

a. 7 quarts = _____ cups

b. 5 miles = _____ inches

c. 3 gallons = _____ fluid ounces

d. 4 days = _____ seconds

Solve It

3. Answer each of the following questions:

    a. If you have 420 minutes, how many hours do you have?

    b. If you have 144 inches, how many yards do you have?

    c. If you have 22,000 pounds, how many tons do you have?

    d. If you have 256 fluid ounces, how many gallons do you have?

Solve It

4. Calculate each of the following:

    a. 168 inches = _____ feet

    b. 100 quarts = _____ gallons

    c. 288 ounces = _____ pounds

    d. 76 cups = _____ quarts

Solve It

Going International with the Metric System

The metric system is the most commonly used system of measurement throughout the world. Scientists and others who like to keep up with the latest lingo (since the 1960s) often refer to it as the *International System of Units,* or *SI.* Unlike the English system, the metric system is based exclusively on powers of ten (see Chapter 10). This feature makes the metric system much easier to use (after you get the hang of it!) than the English system because you can do a lot of calculations simply by moving the decimal point.

The metric system includes five basic units, shown in Table 11-2.

Table 11-2	Five Basic Metric Units
Measure of	*Metric Units*
Distance (length)	Meters (m)
Fluid volume (capacity)	Liters (L)
Mass (weight)	Grams (g)
Time	Seconds (s)
Temperature	Degrees Celsius/Centigrade (°C)

You can modify each basic metric unit with the prefixes shown in Table 11-3. When you know how the metric prefixes work, you can use them to make sense even of units that you're not familiar with.

Table 11-3	Metric Prefixes		
Prefix	**Meaning**	**Number**	**Power of Ten**
Tera-	One trillion	1,000,000,000,000	10^{12}
Giga-	One billion	1,000,000,000	10^{9}
Mega-	One million	1,000,000	10^{6}
Kilo-	One thousand	1,000	10^{3}
Hecta-	One hundred	100	10^{2}
Deca-	Ten	10	10^{1}
(none)	One	1	10^{0}
Deci-	One tenth	0.1	10^{-1}
Centi-	One hundredth	0.01	10^{-2}
Milli-	One thousandth	0.001	10^{-3}
Micro-	One millionth	0.000001	10^{-6}
Nano-	One billionth	0.000000001	10^{-9}

Q. How many millimeters are in a meter?

A. **1,000.** The prefix *milli-* means one *thousandth,* so a millimeter is $\frac{1}{1,000}$ of a meter. Therefore, a meter contains 1,000 millimeters.

Q. A *dyne* is an old unit of force, the push or pull on an object. Using what you know about metric prefixes, how many dynes do you think are in 14 teradynes?

A. **14,000,000,000,000 (14 trillion).** The prefix *tera-* means *one trillion,* so 1,000,000,000,000 dynes are in a teradyne; therefore,

14 teradynes = 14 × 1,000,000,000,000 dynes = 14,000,000,000,000 dynes

5. Give the basic metric unit for each type of measurement listed below:

a. The amount of vegetable oil for a recipe

b. The weight of an elephant

c. How much water a swimming pool can hold

d. How hot a swimming pool is

e. How long you can hold your breath

f. Your height

g. Your weight

h. How far you can run

Solve It

6. Write down the number or decimal associated with each of the following metric prefixes:

a. kilo-

b. milli-

c. centi-

d. mega-

e. micro-

f. giga-

g. nano-

h. no prefix

Solve It

7. Answer each of the following questions:

a. How many centimeters are in a meter?

b. How many milliliters are in a liter?

c. How many milligrams are in a kilogram?

d. How many centimeters are in a kilometer?

Solve It

8. Using what you know about metric prefixes, calculate each of the following:

a. 75 kilowatts = _____ watts

b. 12 seconds = _____ microseconds

c. 7 megatons = _____ tons

d. 400 gigaHertz = _____ Hertz

Solve It

Converting Between English and Metric Units

To convert between metric units and English units, use the four conversion equations shown in the first column of Table 11-4.

Table 11-4	Conversion Factors for English and Metric Units	
Conversion Equation	**English-to-Metric**	**Metric-to-English**
1 meter ≈ 3.26 feet	$\dfrac{1\,m}{3.26\,ft}$	$\dfrac{3.26\,ft}{1\,m}$
1 kilometer ≈ 0.62 miles	$\dfrac{1\,km}{0.62\,mi.}$	$\dfrac{0.62\,mi.}{1\,km}$
1 liter ≈ 0.26 gallons	$\dfrac{1\,L}{0.26\,gal.}$	$\dfrac{0.26\,gal.}{1\,L}$
1 kilogram ≈ 2.20 pounds	$\dfrac{1\,kg}{2.20\,lb.}$	$\dfrac{2.20\,lb.}{1\,kg}$

The remaining columns in Table 11-4 show conversion factors (fractions) that you multiply by to convert from metric units to English or from English units to metric. To convert from one unit to another, multiply by the conversion factor and cancel out any unit that appears in both the numerator and denominator.

Always use the conversion factor that has units you're converting *from* in the *denominator*. For example, to convert from miles to kilometers, use the conversion factor that has miles in the denominator: that is, $\dfrac{1\,km}{0.62\,mi.}$.

Sometimes, you may want to convert between units for which no direct conversion factor exists. In these cases, set up a *conversion chain* to convert via one or more intermediate units. For instance, to convert centimeters into inches, you may go from centimeters to meters to feet to inches. When the conversion chain is set up correctly, every unit cancels out except for the unit that you're converting to. You can set up a conversion chain of any length to solve a problem.

With a long conversion chain, it's sometimes helpful to take an extra step and turn the whole chain into a single fraction. Place all the numerators above a single fraction bar and all the denominators below it, keeping multiplication signs between the numbers.

A chain can include conversion factors built from any conversion equation in this chapter. For example, you know from "The Basics of the English System" that 2 pints = 1 quart, so you can use the following two fractions:

$$\frac{2\,pt.}{1\,qt.} \qquad \frac{1\,qt.}{2\,pt.}$$

Similarly, you know from "Going International with the Metric System" that 1 kilogram = 1,000 grams, so you can use these two fractions:

$$\frac{1\,kg}{1,000\,g} \qquad \frac{1,000\,g}{1\,kg}$$

Q. Convert 5 kilometers to miles.

A. **3.1 miles.** To convert *from* kilometers, multiply 5 kilometers by the conversion factor with kilometers in the denominator:

$$5 \text{ km} \times \frac{0.62 \text{ mi.}}{1 \text{ km}}$$

Now you can cancel the unit *kilometer* in both the numerator and denominator:

$$= 5 \text{ } \cancel{\text{km}} \times \frac{0.62 \text{ mi.}}{1 \text{ } \cancel{\text{km}}}$$

Calculate the result:

$$= 5 \times \frac{0.62 \text{ mi.}}{1} = 3.1 \text{ mi.}$$

Notice that when you set up the conversion correctly, you don't have to think about the unit — it changes from kilometers to miles automatically!

Q. Convert 21 grams to pounds.

A. **0.0462 pounds.** You don't have a conversion factor to convert grams to pounds directly, so set up a conversion chain that makes the following conversion:

grams → kilograms → pounds

To convert grams to kilograms, use the equation 1,000 g = 1 kg. Multiply by the fraction with kilograms in the numerator and grams in the denominator:

$$21 \text{ g} \times \frac{1 \text{ kg}}{1,000 \text{ g}}$$

To convert kilograms to pounds, use the equation 1 kg = 2.2 lb. and multiply by the fraction with pounds in the numerator and kilograms in the denominator:

$$= 21 \text{ g} \times \frac{1 \text{ kg}}{1,000 \text{ g}} \times \frac{2.2 \text{ lb.}}{1 \text{ kg}}$$

Cancel *grams* and *kilograms* in both the numerator and denominator:

$$= 21 \text{ } \cancel{\text{g}} \times \frac{1 \text{ } \cancel{\text{kg}}}{1,000 \text{ } \cancel{\text{g}}} \times \frac{2.2 \text{ lb.}}{1 \text{ } \cancel{\text{kg}}}$$

Finally, calculate the result:

$$= 21 \times \frac{1}{1,000} \times \frac{2.2 \text{ lb.}}{1}$$

$$= 0.021 \times 2.2 \text{ lbs} = 0.0462 \text{ lb.}$$

After doing the conversion, step back and apply a reasonability test to your answers. Does each answer make sense? For example, when you convert grams to kilograms, the number you end up with should be a lot smaller than the number you started with because each kilogram contains lots of grams.

---

9. Convert 8 kilometers to miles.

Solve It

10. If you weigh 72 kilograms, what's your weight in pounds to the nearest whole pound?

Solve It

---

11. If you're 1.8 meters tall, what's your height in inches to the nearest whole inch?

Solve It

12. Change 100 cups to liters, rounded to the nearest whole liter.

Solve It

Answers to Problems in Weighty Questions on Weights and Measures

The following are the answers to the practice questions presented in this chapter.

1 All these questions ask you to convert a large unit to a smaller one, so use multiplication.

a. How many inches are in a yard? **36 inches.** 12 inches are in a foot and 3 feet are in a yard, so

$3 \times 12 = 36$ in.

b. How many hours are in a week? **168 hours.** 24 hours are in a day and 7 days are in a week, so

$24 \times 7 = 168$ hours

c. How many ounces are in a ton? **32,000 ounces.** 16 ounces are in a pound and 2,000 pounds are in a ton, so

$16 \times 2,000 = 32,000$ oz.

d. How many cups are in a gallon? **16 cups.** 2 cups are in a pint, 2 pints are in a quart, and 4 quarts are in a gallon, so

$2 \times 2 \times 4 = 16$ c.

2 All these problems ask you to convert a large unit to a smaller one, so use multiplication:

a. 7 quarts = **28 cups.** 2 cups are in a pint and 2 pints are in a quart, so

$2 \times 2 = 4$ c.

So 4 cups are in a quart; therefore,

7 qt. = 7×4 c. = 28 c.

b. 5 miles = **316, 800 inches.** 12 inches are in a foot and 5,280 feet are in a mile, so

$12 \times 5,280 = 63,360$ in.

63,360 inches are in a mile, so

5 mi. = $5 \times 63,360$ in. = 316,800 in.

c. 3 gallons = **384 fluid ounces.** 8 fluid ounces are in a cup, 2 cups are in a pint, 2 pints are in a quart, and 4 quarts are in a gallon, so

$8 \times 2 \times 2 \times 4 = 128$ fl. oz.

Therefore, 128 fluid ounces are in a gallon, so

3 gal. = 3×128 fl. oz. = 384 fl. oz.

d. 4 days = **345,600 seconds.** 60 seconds are in a minute, 60 minutes are in an hour, and 24 hours are in a day, so

$60 \times 60 \times 24 = 86,400$ seconds

86,400 seconds are in a day, so

4 days = $4 \times 86,400$ seconds = 345,600 seconds

3 All these problems ask you to convert a smaller unit to a larger one, so use division:

a. If you have 420 minutes, how many hours do you have? **7 hours.** There are 60 minutes in an hour, so divide by 60:

420 minutes ÷ 60 = 7 hours

b. If you have 144 inches, how many yards do you have? **4 yards.** There are 12 inches in a foot, so divide by 12:

144 in. ÷ 12 = 12 ft.

There are 3 feet in a yard, so divide by 3:

12 ft. ÷ 3 = 4 yd.

c. If you have 22,000 pounds, how many tons do you have? **11 tons.** There are 2,000 pounds in a ton, so divide by 2,000

22,000 lb. ÷ 2,000 = 11 tons

d. If you have 256 fluid ounces, how many gallons do you have? **2 gallons.** There are 8 fluid ounces in a cup, so divide by 8:

256 fl. oz. ÷ 8 = 32 c.

There are 2 cups in a pint, so divide by 2:

32 c. ÷ 2 = 16 pt.

There are 2 pints in a quart, so divide by 2:

16 pt. ÷ 2 = 8 qt.

There are 4 quarts in a gallon, so divide by 4:

8 qt. ÷ 4 = 2 gal.

4 All these problems ask you to convert a smaller unit to a larger one, so use division:

a. 168 inches = **14 feet.** There are 12 inches in a foot, so divide by 12:

168 in. ÷ 12 = 14 ft.

b. 100 quarts = **25 gallons.** There are 4 quarts in a gallon, so divide by 4:

100 qt. ÷ 4 = 25 gal.

c. 288 ounces = **18 pounds.** There are 16 ounces in a pound, so divide by 16:

288 oz. ÷ 16 = 18 lbs.

d. 76 cups = **19 quarts.** There are 2 cups in a pint, so divide by 2:

76 cups ÷ 2 = 38 pints

There are 2 pints in a quart, so divide by 2:

38 pints ÷ 2 = 19 quarts

5 Give the *basic metric unit* for each type of measurement listed below. Note that you want only the base unit, without any prefixes.

a. The amount of vegetable oil for a recipe: **liters**

b. The weight of an elephant: **grams**

c. How much water a swimming pool can hold: **liters**

  d. How hot a swimming pool is: **degrees Celsius (Centigrade)**

  e. How long you can hold your breath: **seconds**

  f. Your height: **meters**

  g. Your weight: **grams**

  h. How far you can run: **meters**

6 Write down the number or decimal associated with each of the following metric prefixes:

  a. kilo-: **1,000 (one thousand or 10^3)**

  b. milli-: **0.001 (one thousandth or 10^{-3})**

  c. centi-: **.01 (one hundredth or 10^{-2})**

  d. mega-: **1,000,000 (one million or 10^6)**

  e. micro-: **0.000001 (one millionth or 10^{-6})**

  f. giga-: **1,000,000,000 (one billion or 10^9)**

  g. nano-: **0.000000001 (one billionth or 10^{-9})**

  h. no prefix: **1 (one or 10^0)**

7 Answer each of the following questions:

  a. How many centimeters are in a meter? **100 centimeters**

  b. How many milliliters are in a liter? **1,000 milliliters**

  c. How many milligrams are in a kilogram? **1,000,000 milligrams.** 1,000 milligrams are in a gram and 1,000 grams are in a kilogram, so

  $1,000 \times 1,000 = 1,000,000$ mg

  Therefore, 1,000,000 milligrams are in a kilogram.

  d. How many centimeters are in a kilometer? **100,000 centimeters.** 100 centimeters are in a meter and 1,000 meters are in a kilometer, so

  $100 \times 1,000 = 100,000$ cm

  Therefore, 100,000 centimeters are in a kilometer.

8 Using what you know about metric prefixes, calculate each of the following:

  a. 75 kilowatts = **75,000 watts.** The prefix *kilo-* means *one thousand,* so 1,000 watts are in a kilowatt; therefore,

  75 kilowatts = $75 \times 1,000$ watts = 75,000 watts

  b. 12 seconds = **12,000,000 microseconds.** The prefix *micro-* means *one millionth,* so a microsecond is a millionth of a second. Therefore, 1,000,000 microseconds are in a second. Thus,

  12 seconds = $12 \times 1,000,000$ microseconds = 12,000,000 microseconds

  c. 7 megatons = **7,000,000 tons.** The prefix *mega-* means *one million,* so 1,000,000 tons are in a megaton; therefore,

  7 megatons = $7 \times 1,000,000$ tons = 7,000,000 tons

  d. 400 gigaHertz = **400,000,000,000 Hertz.** The prefix *giga-* means *one billion,* so 1,000,000,000 Hertz are in a gigaHertz; thus,

  400 gigaHertz = $400 \times 1,000,000,000$ Hertz = 400,000,000,000 gigaHertz

9   8 kilometers = **4.96 miles.** To convert kilometers to miles, multiply by the conversion fraction with miles in the numerator and kilometers in the denominator:

$$8 \text{ km} \times \frac{0.62 \text{ mi.}}{1 \text{ km}}$$

Cancel the unit *kilometer* in both the numerator and denominator:

$$= 8 \text{ } \cancel{\text{km}} \times \frac{0.62 \text{ mi.}}{1 \text{ } \cancel{\text{km}}}$$

Now calculate the result:

$$= 8 \times 0.62 \text{ mi.} = 4.96 \text{ mi.}$$

10   To the nearest pound, 72 kilograms = **158 pounds.** To convert kilograms to pounds, multiply by the conversion factor with pounds in the numerator and kilograms in the denominator:

$$72 \text{ kg} \times \frac{2.2 \text{ lb.}}{1 \text{ kg}}$$

Cancel the unit *kilogram* in both the numerator and denominator:

$$= 72 \text{ } \cancel{\text{kg}} \times \frac{2.2 \text{ lb.}}{1 \text{ } \cancel{\text{kg}}}$$

$$= 72 \times 2.2 \text{ lb.}$$

Multiply to find the answer:

$$= 158.4 \text{ lb.}$$

Round to the nearest whole pound:

$$\approx 158 \text{ lb.}$$

11   To the nearest inch, 1.8 meters = **70 inches.** You don't have a conversion factor to change meters to inches directly, so set up a conversion chain as follows:

meters → feet → inches

Convert meters to feet with the fraction that puts meters in the denominator:

$$1.8 \text{ m} \times \frac{3.26 \text{ ft.}}{1 \text{ m}}$$

Convert the feet to inches with the conversion factor that has feet in the denominator:

$$= 1.8 \text{ m} \times \frac{3.26 \text{ ft.}}{1 \text{ m}} \times \frac{12 \text{ in.}}{1 \text{ ft.}}$$

Cancel the units *meters* and *feet* in both the numerator and denominator:

$$= 1.8 \text{ } \cancel{\text{m}} \times \frac{3.26 \text{ } \cancel{\text{ft.}}}{1 \text{ } \cancel{\text{m}}} \times \frac{12 \text{ in.}}{1 \text{ } \cancel{\text{ft.}}}$$

Multiply to find the answer:

$$= 70.416 \text{ in.}$$

Round the answer to the nearest whole inch:

$$\approx 70 \text{ in.}$$

12 To the nearest liter, 100 cups = **24 liters.** You don't have a conversion factor to change cups to liters directly, so set up a conversion chain:

cups → pints → quarts → gallons → liters

Convert cups to pints:

$$100 \text{ c.} \times \frac{1 \text{ pt.}}{2 \text{ c.}}$$

Convert pints to quarts:

$$= 100 \text{ c.} \times \frac{1 \text{ pt.}}{2 \text{ c.}} \times \frac{1 \text{ qt.}}{2 \text{ pt.}}$$

Convert quarts to gallons:

$$= 100 \text{ c.} \times \frac{1 \text{ pt.}}{2 \text{ c.}} \times \frac{1 \text{ qt.}}{2 \text{ pt.}} \times \frac{1 \text{ gal.}}{4 \text{ qt.}}$$

Convert gallons to liters:

$$= 100 \text{ c.} \times \frac{1 \text{ pt.}}{2 \text{ c.}} \times \frac{1 \text{ qt.}}{2 \text{ pt.}} \times \frac{1 \text{ gal.}}{4 \text{ qt.}} \times \frac{1 \text{ L}}{0.26 \text{ gal.}}$$

Now all units *except* liters cancel out:

$$= 100 \, \cancel{\text{c.}} \times \frac{1 \, \cancel{\text{pt.}}}{2 \, \cancel{\text{c.}}} \times \frac{1 \, \cancel{\text{qt.}}}{2 \, \cancel{\text{pt.}}} \times \frac{1 \, \cancel{\text{gal.}}}{4 \, \cancel{\text{qt.}}} \times \frac{1 \text{ L}}{0.26 \, \cancel{\text{gal.}}}$$

$$= 100 \times \frac{1}{2} \times \frac{1}{2} \times \frac{1}{4} \times \frac{1 \text{ L}}{0.26}$$

To avoid confusion, set this chain up as a single fraction:

$$\frac{100 \times 1 \times 1 \times 1 \times 1 \text{ L}}{2 \times 2 \times 4 \times 0.26}$$

$$= \frac{100 \text{ L}}{4.16}$$

Use decimal division to find the answer to at least one decimal place:

$$\approx 24.04 \text{ L}$$

Round to the nearest whole liter:

$$\approx 24 \text{ L}$$

Chapter 12

Shaping Up with Geometry

- -

In This Chapter

▶ Measuring quadrilaterals

▶ Finding the area of a triangle

▶ Using the Pythagorean theorem

▶ Finding the area and circumference of a circle

▶ Calculating the volume of a variety of solids

- -

Geometry is the study of shapes and figures, and it's really popular with the Ancient Greeks, architects, engineers, carpenters, robot designers, and high school math teachers.

A *shape* is any closed two-dimensional (2-D) geometric figure that has an inside and an outside, and a *solid* is just like a shape, only it's three-dimensional (in 3-D). In this chapter, you work with three important shapes: quadrilaterals, triangles, and circles. I show you how to find the area and in some cases the *perimeter* (the length of the edge) of these shapes. I also focus on a variety of solids, showing you how to find the volume of each.

Getting in Shape: Polygon (And Non-Polygon) Basics

You can divide shapes into two basic types: polygons and non-polygons. A *polygon* has all straight sides, and you can easily classify polygons by the number of sides they have:

Polygon	Number of Sides
Triangle	3
Quadrilateral	4
Pentagon	5
Hexagon	6
Heptagon	7
Octagon	8

Any shape that has at least one curved edge is a *non-polygon*. The most common non-polygon is the circle.

REMEMBER

The area of a shape — the space inside — is usually measured in square units, such as square inches (in.²), square meters (m²), or square kilometers (km²). If a problem mixes units of measurement, such as inches and feet, you have to convert to one or the other before doing the math (for more on conversions, see Chapter 11).

Squaring Off with Quadrilaterals

Any shape with four sides is a *quadrilateral*. Quadrilaterals include squares, rectangles, rhombuses, parallelograms, and trapezoids, plus a host of more irregular shapes. In this section, I show you how to find the area (*A*) and in some cases the perimeter (*P*) of these five basic types of quadrilaterals.

A *square* has four right angles and four equal sides. To find the area and perimeter of a square, use the following formulas, where *s* stands for the length of a side (see Figure 12-1):

$$A = s^2$$
$$P = 4 \times s$$

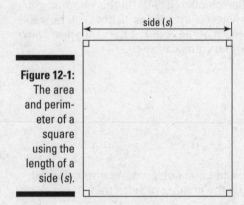

Square

side (*s*)

Figure 12-1:
The area and perimeter of a square using the length of a side (*s*).

A *rectangle* has four right angles and opposite sides that are equal. The long side of a rectangle is called the *length,* and the short side is called the *width*. To find the area and perimeter of a rectangle, use the following formulas, where *l* stands for the length of a side and *w* stands for width (see Figure 12-2):

$$A = l \times w$$
$$P = 2 \times (l + w)$$

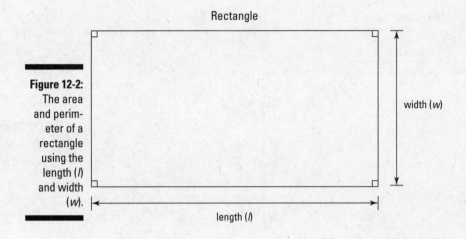

Figure 12-2:
The area
and perim-
eter of a
rectangle
using the
length (*l*)
and width
(*w*).

A *rhombus* resembles a collapsed square. It has four equal sides, but its four angles aren't necessarily right angles. Similarly, a *parallelogram* resembles a collapsed rectangle. Its opposite sides are equal, but its four angles aren't necessarily right angles. To find the area of a rhombus or a parallelogram, use the following formula, where *b* stands for the length of the base (either the bottom or top side) and *h* stands for the height (the shortest distance between the two bases); also see Figure 12-3:

$$A = b \times h$$

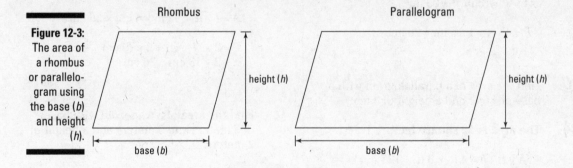

Figure 12-3:
The area of
a rhombus
or parallelo-
gram using
the base (*b*)
and height
(*h*).

A *trapezoid* is a quadrilateral whose only distinguishing feature is that it has two parallel bases (top side and bottom side). To find the area of a trapezoid, use the following formula, where b_1 and b_2 stand for the lengths of the two bases and *h* stands for the height (the shortest distance between the two bases); also see Figure 12-4:

$$A = \frac{1}{2} \times (b_1 + b_2) \times h$$

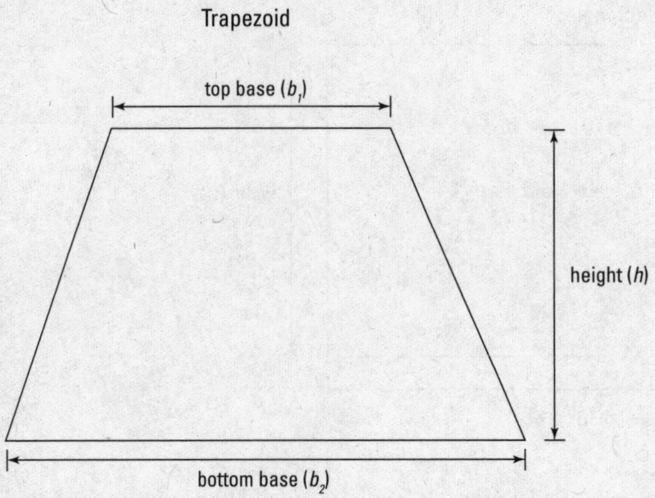

Trapezoid

top base (b_1)

height (h)

bottom base (b_2)

Figure 12-4:
The area of a trapezoid using the length of the two bases (b_1 and b_2) and the height (h).

EXAMPLE

Q. Find the area and perimeter of a square with a side that measures 5 inches.

A. **The area is 25 square inches, and the perimeter is 20 inches.**

$$A = s^2 = (5 \text{ in.})^2 = 25 \text{ in.}^2$$

$$P = 4 \times s = 4 \times 5 \text{ in.} = 20 \text{ in.}$$

Q. Find the area of a parallelogram with a base of 4 feet and a height of 3 feet.

A. **The area is 12 square feet.**

$$A = b \times h = 4 \text{ ft.} \times 3 \text{ ft.} = 12 \text{ ft.}^2$$

Q. Find the area and perimeter of a rectangle with a length of 9 centimeters and a width of 4 centimeters.

A. **The area is 36 square centimeters, and the perimeter is 26 centimeters.**

$$A = l \times w = 9 \text{ cm} \times 4 \text{ cm} = 36 \text{ cm}^2$$

$$P = 2 \times (l + w) = 2 \times (9 \text{ cm} + 4 \text{ cm})$$
$$= 2 \times 13 \text{ cm} = 26 \text{ cm}$$

Q. Find the area of a trapezoid with bases of 3 meters and 5 meters and a height of 2 meters.

A. **The area is 8 square meters.**

$$A = \frac{1}{2} \times (b_1 + b_2) \times h$$
$$= \frac{1}{2} \times (3 \text{ m} + 5 \text{ m}) \times 2 \text{ m}$$
$$= \frac{1}{2} \times (8 \text{ m}) \times 2 \text{ m} = 8 \text{ m}^2$$

1. What are the area and perimeter of a square with a side of 9 miles?

Solve It

2. Find the area and perimeter of a square with a side of 31 centimeters.

Solve It

3. Figure out the area and perimeter of a rectangle with a length of 10 inches and a width of 5 inches.

Solve It

4. Determine the area and perimeter of a rectangle that has a length of 23 kilometers and a width of 19 kilometers.

Solve It

5. What's the area of a rhombus with a base of 9 meters and a height of 6 meters?

Solve It

6. Figure out the area of a parallelogram with a base of 17 yards and a height of 13 yards.

Solve It

7. Write down the area of a trapezoid with bases of 6 feet and 8 feet and a height of 5 feet.

Solve It

8. What's the area of a trapezoid that has bases of 15 millimeters and 35 millimeters and a height of 21 millimeters?

Solve It

Making a Triple Play with Triangles

Any shape with three straight sides is a *triangle*. To find the area of a triangle, use the following formula, in which *b* is the length of the base (one side of the triangle) and *h* is the height of the triangle (the shortest distance from the base to the opposite corner); also see Figure 12-5:

$$A = \frac{1}{2} \times b \times h$$

Triangle

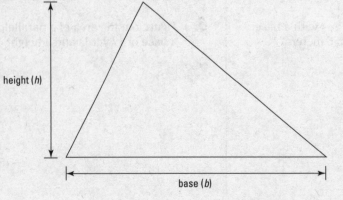

Figure 12-5:
The area of a triangle using the base (*b*) and height (*h*).

height (*h*)

base (*b*)

Any triangle that has a 90-degree angle is called a *right triangle*. The right triangle is one of the most important shapes in geometry. In fact, *trigonometry*, which is devoted entirely to the study of triangles, begins with a set of key insights and observations about right triangles.

The longest side of a right triangle (c) is called the *hypotenuse,* and the two short sides (a and b) are each called *legs.* The most important right triangle formula — called the *Pythagorean theorem* — allows you to find the length of the hypotenuse given only the length of the legs:

$$a^2 + b^2 = c^2$$

Figure 12-6 shows this theorem in action.

Right Triangle

Figure 12-6: Using the Pythagorean theorem to find the hypotenuse (c) of a right triangle.

leg (a)

hypotenuse (c)

leg (b)

EXAMPLE

Q. Find the area of a triangle with a base of 5 meters and a height of 6 meters.

A. **15 square meters.**

$$A = \frac{1}{2} \times 5 \text{ m} \times 6 \text{ m} = 15 \text{ m}^2$$

Q. Find the hypotenuse of a right triangle with legs that are 6 inches and 8 inches in length.

A. **10 inches.** Use the Pythagorean theorem to find the value of c as follows:

$$a^2 + b^2 = c^2$$

$$6^2 + 8^2 = c^2$$

$$36 + 64 = c^2$$

$$100 = c^2$$

So when you multiply c by itself, the result is 100. Therefore, $c = 10$ in., because $10 \times 10 = 100$.

9. What's the area of a triangle with a base of 7 centimeters and a height of 4 centimeters?

Solve It

10. Find the area of a triangle with a base of 10 kilometers and a height of 17 kilometers.

Solve It

11. Figure out the area of a triangle with a base of 2 feet and a height of 33 inches.

Solve It

12. Discover the hypotenuse of a right triangle whose two legs measure 3 miles and 4 miles.

Solve It

13. What's the hypotenuse of a right triangle with two legs measuring 5 millimeters and 12 millimeters?

Solve It

14. Calculate the hypotenuse of a right triangle with two legs measuring 8 feet and 15 feet.

Solve It

Getting Around with Circle Measurements

A *circle* is the set of all points that are a constant distance from a point inside it. Here are a few terms that are handy when talking about circles (see Figure 12-7):

- ✔ The *center* (*c*) of a circle is the point that's the same distance from any point on the circle itself.
- ✔ The *radius* (*r*) of a circle is the distance from the center to any point on the circle.
- ✔ The *diameter* (*d*) of a circle is the distance from any point on the circle through the center to the opposite point on the circle.

Circle

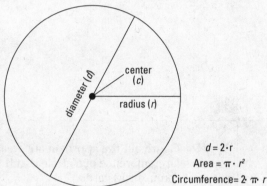

Figure 12-7:
The area and circumference of a circle using the radius (*r*).

$d = 2 \cdot r$

Area $= \pi \cdot r^2$

Circumference $= 2 \cdot \pi \cdot r$

To find the area (*A*) of a circle, use the following formula:

$$A = \pi \times r^2$$

The symbol π is called *pi* (pronounced *pie*). It's a decimal that goes on forever, so you can't get an exact value for π. However, the number 3.14 is a good approximation of π that you can use when solving problems that involve circles. (Note that when you use an approximation, the $\approx$ symbol replaces the = sign in problems.)

The perimeter of a circle has a special name: the *circumference* (*C*). The formulas for the circumference of a circle also include π:

$$C = 2 \times \pi \times r$$

$$C = \pi \times d$$

These circumference formulas say the same thing because, as you can see in Figure 12-7, the diameter of a circle is always twice the radius of that circle. That gives you the following formula:

$$d = 2 \times r$$

Q. What's the diameter of a circle that has a radius of 3 inches?

A. **6 inches.**

$$d = 2 \times r = 2 \times 3 \text{ in.} = 6 \text{ in.}$$

Q. What's the approximate area of a circle that has a radius of 10 millimeters?

A. **314 square millimeters.**

$$A = \pi \times r^2$$

$$\approx 3.14 \times (10 \text{ mm})^2$$

$$\approx 3.14 \times 100 \text{ mm}^2$$

$$\approx 314 \text{ mm}^2$$

Q. What's the approximate circumference of a circle that has a radius of 4 feet?

A. **25.12 feet.**

$$C = 2\pi \times r$$

$$\approx 2 \times 3.14 \times 4 \text{ ft.}$$

$$\approx 25.12 \text{ ft.}$$

15. What's the approximate area and circumference of a circle that has a radius of 3 kilometers?

Solve It

16. Figure out the approximate area and circumference of a circle that has a radius of 12 yards.

Solve It

17. Write down the approximate area and circumference of a circle with a diameter of 52 centimeters.

Solve It

18. Find the approximate area and circumference of a circle that has a diameter of 86 inches.

Solve It

Building Solid Measurement Skills

Solids take you into the real world, the third dimension. One of the simplest solids is the *cube*, a solid with six identical square faces. To find the volume of a cube, use the following formula, where *s* is the length of the side of any one face (check out Figure 12-8):

$$V = s^3$$

A *box* (also called a *rectangular solid*) is a solid with six rectangular faces. To find the volume of a box, use the following formula, where *l* is the length, *w* is the width, and *h* is the height (see Figure 12-9):

$$V = l \times w \times h$$

Cube

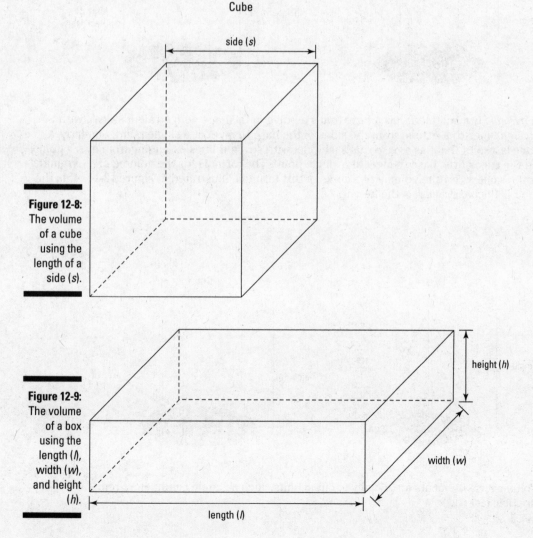

Figure 12-8:
The volume of a cube using the length of a side (*s*).

side (*s*)

Figure 12-9:
The volume of a box using the length (*l*), width (*w*), and height (*h*).

height (*h*)

width (*w*)

length (*l*)

A *prism* is a solid with two identical bases and a constant cross-section — that is, whenever you slice a prism parallel to the bases, the cross-section is the same size and shape as the bases. A *cylinder* is a solid with two identical circular bases and a constant cross-section. To find the volume of a prism or cylinder, use the following formula, where A_b is the area of the base and h is the height (see Figure 12-10). You can find the area formulas throughout this chapter:

$$V = A_b \times h$$

Figure 12-10:
The volume of a prism or cylinder using the area of the base (A_b) and the height (h).

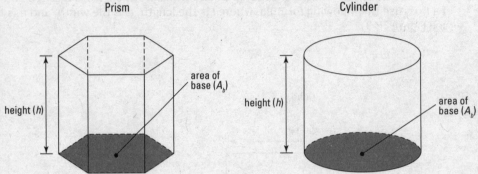

A *pyramid* is a solid that has a base that's a polygon (a shape with straight sides), with straight lines that extend from the sides of the base to meet at a single point. Similarly, a *cone* is a solid that has a base that's a circle, with straight lines extending from every point on the edge of the base to meet at a single point. The formula for the volume of a pyramid is the same as for the volume of a cone. In this formula, illustrated in Figure 12-11, A_b is the area of the base, and h is the height:

$$V = \frac{1}{3} \times A_b \times h$$

Figure 12-11:
The volume of a pyramid or cone using the area of the base (A_b) and the height (h).

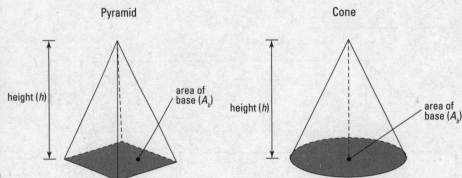

Volume measurements are usually in cubic units, such as cubic centimeters (cm^3) or cubic feet ($ft.^3$).

Q. What's the volume of a cube with a side that measures 4 centimeters?

A. **64 cubic centimeters.**

$$V = s^3 = (4 \text{ cm})^3 = 4 \text{ cm} \times 4 \text{ cm} \times 4 \text{ cm}$$
$$= 64 \text{ cm}^3$$

Q. Calculate the volume of a box with a length of 7 inches, a width of 4 inches, and a height of 2 inches.

A. **56 cubic inches.**

$$V = l \times w \times h = 7 \text{ in.} \times 4 \text{ in.} \times 2 \text{ in.} = 56 \text{ in.}^3$$

Q. Find the volume of a prism with a base that has an area of 6 square centimeters and a height of 3 centimeters.

A. **18 cubic centimeters.**

$$V = A_b \times h = 6 \text{ cm}^2 \times 3 \text{ cm} = 18 \text{ cm}^3$$

Q. Figure out the approximate volume of a cylinder with a base that has a radius of 2 feet and a height of 8 feet.

A. **100.48 cubic feet.** To begin, find the approximate area of the base using the formula for the area of a circle:

$$A_b = \pi \times r^2$$
$$\approx 3.14 \times (2 \text{ ft.})^2$$
$$\approx 3.14 \times 4 \text{ ft.}^2$$
$$\approx 12.56 \text{ ft.}^2$$

Now plug this result into the formula for the volume of a prism/cylinder:

$$V = A_b \times h$$
$$= 12.56 \text{ ft.}^2 \times 8 \text{ ft.}$$
$$= 100.48 \text{ ft.}^3$$

Q. Find the volume of a pyramid with a square base whose side is 10 inches and with a height of 6 inches.

A. **200 cubic inches.** First, find the area of the base using the formula for the area of a square:

$$A_b = s^2 = (10 \text{ in.})^2 = 100 \text{ in.}^2$$

Now plug this result into the formula for the volume of a pyramid/cone:

$$V = \frac{1}{3} \times A_b \times h$$
$$= \frac{1}{3} \times 100 \text{ in.}^2 \times 6 \text{ in.}$$
$$= 200 \text{ in.}^3$$

Q. Find the approximate volume of a cone with a base that has a radius of 2 meters and with a height of 3 meters.

A. **12.56 cubic meters.** First, find the approximate area of the base using the formula for the area of a circle:

$$A_b = \pi \times r^2$$
$$\approx 3.14 \times (2 \text{ m})^2$$
$$\approx 3.14 \times 4 \text{ m}^2$$
$$= 12.56 \text{ m}^2$$

Now plug this result into the formula for the volume of a pyramid/cone:

$$V = \frac{1}{3} \times A_b \times h$$
$$= \frac{1}{3} \times 12.56 \text{ m}^2 \times 3 \text{ m}$$
$$= 12.56 \text{ m}^3$$

19. Find the volume of a cube that has a side of 19 meters.

Solve It

20. Figure out the volume of a box with a length of 18 centimeters, a width of 14 centimeters, and a height of 10 centimeters.

Solve It

21. Figure out the approximate volume of a cylinder whose base has a radius of 7 millimeters and whose height is 16 millimeters.

Solve It

22. Find the approximate volume of a cone whose base has a radius of 3 inches and whose height is 8 inches.

Solve It

Answers to Problems in Shaping Up with Geometry

The following are the answers to the practice questions presented in this chapter.

1 **Area is 81 square miles; perimeter is 36 miles.** Use the formulas for the area and perimeter of a square:

$$A = s^2 = (9 \text{ mi.})^2 = 81 \text{ mi.}^2$$

$$P = 4 \times s = 4 \times 9 \text{ mi.} = 36 \text{ mi.}$$

2 **Area is 961 square centimeters; perimeter is 124 centimeters.** Plug in 31 cm for s in the formulas for the area and perimeter of a square.

$$A = s^2 = (31 \text{ cm})^2 = 961 \text{ cm}^2$$

$$P = 4 \times s = 4 \times 31 \text{ cm} = 124 \text{ cm}$$

3 **Area is 50 square inches; perimeter is 30 inches.** Plug the length and width into the area and perimeter formulas for a rectangle:

$$A = l \times w = 10 \text{ in.} \times 5 \text{ in.} = 50 \text{ in.}^2$$

$$P = 2 \times (l + w) = 2 \times (10 \text{ in} + 5 \text{ in}) = 30 \text{ in}$$

4 **Area is 437 square kilometers; perimeter is 84 kilometers.** Use the rectangle area and perimeter formulas:

$$A = l \times w = 23 \text{ km} \times 19 \text{ km} = 437 \text{ km}^2$$

$$P = 2 \times (l + w) = 2 \times (23 \text{ km} + 19 \text{ km}) = 84 \text{ km}$$

5 **54 square meters.** Use the parallelogram/rhombus area formula:

$$A = b \times h = 9 \text{ m} \times 6 \text{ m} = 54 \text{ m}^2$$

6 **221 square yards.** Use the parallelogram/rhombus area formula:

$$A = b \times h = 17 \text{ yd.} \times 13 \text{ yd.} = 221 \text{ yd.}^2$$

7 **35 square feet.** Plug your numbers into the trapezoid area formula:

$$A = \frac{1}{2} \times (b_1 + b_2) \times h$$

$$= \frac{1}{2} \times (6 \text{ ft.} + 8 \text{ ft.}) \times 5 \text{ ft.}$$

$$= \frac{1}{2} \times 14 \text{ ft.} \times 5 \text{ ft.}$$

$$= 35 \text{ ft.}^2$$

8 **525 square millimeters.** Use the trapezoid area formula:

$$A = \frac{1}{2} \times (b_1 + b_2) \times h$$
$$= \frac{1}{2} \times (15 \text{ mm} + 35 \text{ mm}) \times 21 \text{ mm}$$
$$= \frac{1}{2} \times 50 \text{ mm} \times 21$$
$$= 525 \text{ mm}^2$$

9 **14 square centimeters.** Use the triangle area formula:

$$A = \frac{1}{2} \times b \times h = \frac{1}{2} \times 7 \text{ cm} \times 4 \text{ cm} = 14 \text{ cm}^2$$

10 **85 square kilometers.** Plug in the numbers for the base and height of the triangle:

$$A = \frac{1}{2} \times b \times h = \frac{1}{2} \times 10 \text{ km} \times 17 \text{ km} = 85 \text{ km}^2$$

11 **396 square inches.** First, convert feet to inches. Twelve inches are in 1 foot:

2 ft. = 24 in.

Now use the area formula for a triangle:

$$A = \frac{1}{2} \times b \times h$$
$$= \frac{1}{2} \times 24 \text{ in.} \times 33 \text{ in.}$$
$$= 396 \text{ in.}^2$$

Note: If you instead converted from inches to feet, the answer 2.75 square feet is also correct.

12 **5 miles.** Use the Pythagorean theorem to find the value of c as follows:

$a^2 + b^2 = c^2$

$3^2 + 4^2 = c^2$

$9 + 16 = c^2$

$25 = c^2$

When you multiply c by itself, the result is 25. Therefore,

$c = 5$ mi.

13 **13 millimeters.** Use the Pythagorean theorem to find the value of c:

$a^2 + b^2 = c^2$

$5^2 + 12^2 = c^2$

$25 + 144 = c^2$

$169 = c^2$

When you multiply c by itself, the result is 169. The hypotenuse is longer than the longest leg, so c has to be greater than 12. Use trial and error, starting with 13:

$13^2 = 169$

Therefore, the hypotenuse is 13 mm.

14 **17 feet.** Use the Pythagorean theorem to find the value of c:

$$a^2 + b^2 = c^2$$

$$8^2 + 15^2 = c^2$$

$$64 + 225 = c^2$$

$$289 = c^2$$

When you multiply c by itself, the result is 289. The hypotenuse is longer than the longest leg, so c has to be greater than 15. Use trial and error, starting with 16:

$$16^2 = 256$$

$$17^2 = 289$$

Therefore, the hypotenuse is 17 ft.

15 **Approximate area is 28.26 square kilometers; approximate circumference is 18.84 kilometers.** Use the area formula for a circle to find the area:

$$A = \pi \times r^2$$

$$\approx 3.14 \times (3 \text{ km})^2$$

$$= 3.14 \times 9 \text{ km}^2$$

$$= 28.26 \text{ km}^2$$

Use the circumference formula to find the circumference:

$$C = 2\pi \times r$$

$$\approx 2 \times 3.14 \times 3 \text{ km}$$

$$= 18.84 \text{ km}$$

16 **Approximate area is 452.16 square yards; approximate circumference is 75.36 yards.** Use the area formula for a circle to find the area:

$$A = \pi \times r^2$$

$$\approx 3.14 \times (12 \text{ yd.})^2$$

$$= 3.14 \times 144 \text{ yd.}^2$$

$$= 452.16 \text{ yd.}^2$$

Use the circumference formula to find the circumference:

$$C = 2\pi \times r$$

$$\approx 2 \times 3.14 \times 12 \text{ yd.}$$

$$= 75.36 \text{ yd.}$$

17 **Approximate area is 2,122.64 square centimeters; approximate circumference is 163.28 centimeters.** The diameter is 52 cm, so the radius is half of that, which is 26 cm. Use the area formula for a circle to find the area:

$$A = \pi \times r^2$$

$$\approx 3.14 \times (26 \text{ cm})^2$$

$$= 3.14 \times 676 \text{ cm}^2$$

$$= 2,122.64 \text{ cm}^2$$

Use the circumference formula to find the circumference:

$$C = 2\pi \times r$$

$$\approx 2 \times 3.14 \times 26 \text{ cm}$$

$$= 163.28 \text{ cm}$$

18 **Approximate area is 5,805.86 square inches; approximate circumference is 270.04 inches.** The diameter is 86 in., so the radius is half of that, which is 43 in. Use the area formula for a circle to find the area:

$$A = \pi \times r^2$$

$$\approx 3.14 \times (43 \text{ in.})^2$$

$$= 3.14 \times 1,849 \text{ in.}^2$$

$$= 5,805.86 \text{ in.}^2$$

Use the circumference formula to find the circumference:

$$C = 2\pi \times r$$

$$\approx 2 \times 3.14 \times 43 \text{ in.}$$

$$= 270.04 \text{ in.}$$

19 **6,859 cubic meters.** Substitute 19 m for s in the cube volume formula:

$$V = s^3 = (19 \text{ m})^3 = 6,859 \text{ m}^3$$

20 **2,520 cubic centimeters.** Use the box area formula:

$$V = l \times w \times h$$

$$= 18 \text{ cm} \times 14 \text{ cm} \times 10 \text{ cm} = 2,520 \text{ cm}^3$$

21 **Approximately 2,461.76 cubic millimeters.** First, use the area formula for a circle to find the area of the base:

$$A_b = \pi \times r^2$$

$$\approx 3.14 \times (7 \text{ mm})^2$$

$$= 3.14 \times 49 \text{ mm}^2$$

$$= 153.86 \text{ mm}^2$$

Plug this result into the formula for the volume of a prism/cylinder:

$$V = A_b \times h$$

$$= 153.86 \text{ mm}^2 \times 16 \text{ mm} = 2,461.76 \text{ mm}^3$$

22 **Approximately 75.36 cubic inches.** Use the area formula for a circle to find the area of the base:

$$A_b = \pi \times r^2$$
$$\approx 3.14 \times (3 \text{ in.})^2$$
$$= 3.14 \times 9 \text{ in.}^2$$
$$= 28.26 \text{ in.}^2$$

Plug this result into the formula for the volume of a pyramid/cone:

$$V = \frac{1}{3} \times A_b \times h$$
$$= \frac{1}{3} \times 28.26 \text{ in.}^2 \times 8 \text{ in.} = 75.36 \text{ in.}^3$$

Chapter 13

Getting Graphic: *Xy*-Graphs

● ●

In This Chapter

▶ Understanding the basics of the *xy*-graph

▶ Plotting points and drawing lines on the *xy*-graphs

● ●

A *graph* is a visual tool for providing information about numbers. Graphs commonly appear in business reports, sales brochures, newspapers, and magazines — any place where conveying numerical information quickly and clearly is important.

In this chapter, I show you how to work with the graph that's most commonly used in mathematics: the *xy*-graph.

Getting the Point of the Xy-Graph

In math, the most commonly used graph is the *xy-graph* (also called the *Cartesian coordinate system* or the *Cartesian plane*). The *xy*-graph is basically two number lines that cross at 0. These number lines are called the *x-axis* (which runs horizontally) and the *y-axis* (which runs vertically). These two axes (plural of *axis*) cross at a point called the *origin*.

Every point on the *xy*-graph is represented by a pair of numbers in parentheses, called a set of *xy-coordinates* (or an *ordered pair*). The first number is called the *x-coordinate,* and the second is called the *y-coordinate.*

To *plot* (locate) a point on the *xy*-graph, start at the origin (0,0) and follow the coordinates:

> ✔ **The *x*-coordinate:** The first number tells you how far to go to the right (if positive) or left (if negative) along the *x*-axis.

> ✔ **The *y*-coordinate:** The second number tells you how far to go up (if positive) or down (if negative) along the *y*-axis.

For example, to plot the point $P = (3,-4)$, start at the origin and count three places to the right; then travel down four places and plot your point there.

Q. Plot the following points on the *xy*-graph:

a. $A = (2,5)$

b. $B = (-3,1)$

c. $C = (-2,-4)$

d. $D = (6,0)$

e. $E = (-5,-5)$

f. $F = (0,-1)$

A.

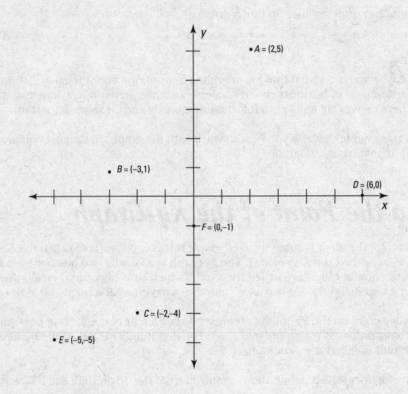

Q. Write down the _xy_-coordinates of points _G_ through _L_.

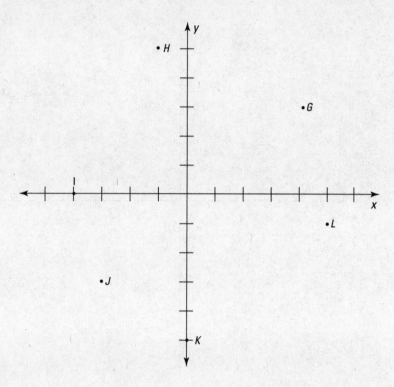

A. _G_ = (4,3), _H_ = (–1,5), _I_ = (–4,0), _J_ = (–3,–3), _K_ = (0,–5), and _L_ = (5,–1).

1. Plot each of the following points on the *xy*-graph.

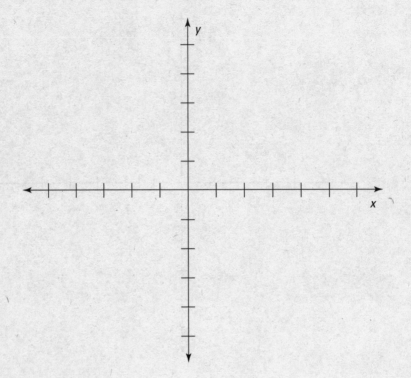

a. $M = (5,6)$

b. $N = (-6,-2)$

c. $O = (0,0)$

d. $P = (-1,2)$

e. $Q = (3,0)$

f. $R = (0,2)$

Solve It

2. Write down the *xy*-coordinates for each point *S* through *X*.

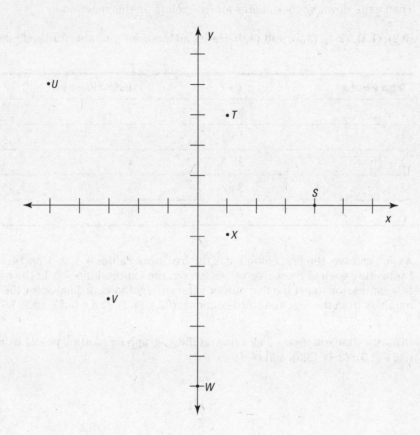

Drawing the Line on the Xy-Graph

After you understand how to plot points on a *xy*-graph (see the preceding section), you can use this skill to draw lines that represent equations on the graph. To see how this works, it's helpful to understand the concept of a function.

A *function* is a mathematical machine — often in the form *y* = some expression that involves *x* — that turns one number into another number. The number you start with is called the *input,* and the new number that comes out is the *output.* On a graph, the input is usually *x,* and the output is usually *y.*

A useful tool for understanding functions is an *input-output table.* In such a table, you plug various *x*-values into your formula and do the necessary calculations to find the corresponding *y*-values.

You can use the *xy*-coordinates from an input-output table to plot points on the graph (as I show you in the preceding section). When these points all line up, draw a straight line through them to represent the function on the graph. ***Note:*** Technically, you need to plot only two points to figure out where the line should go. Still, finding more points is good practice, and it's important when you're graphing a function that isn't a straight line.

Q. Make an input-output table for the function $y = x + 2$ for the input values 0, 1, 2, 3, and 4. Then write down xy-coordinates for five points on this function.

A. **(0,2), (1,3), (2,4), (3,5), and (4,6).** Here's an input-output table for the function $y = x + 2$:

Input Value x	x + 2	Output Value y
0	0 + 2	2
1	1 + 2	3
2	2 + 2	4
3	3 + 2	5
4	4 + 2	6

As you can see, the first column has the five input values 0, 1, 2, 3, and 4. In the second column, I substitute each of these five values for x in the expression $x + 2$. In the third column, I evaluate this expression to get the five output values. To get xy-coordinates for the function, pair up numbers from the first and third columns: (0,2), (1,3), (2,4), (3,5), and (4,6).

Q. Draw the function $y = x + 2$ as a line on the xy-graph by plotting points using the coordinates (0,2), (1,3), (2,4), (3,5), and (4,6).

A.

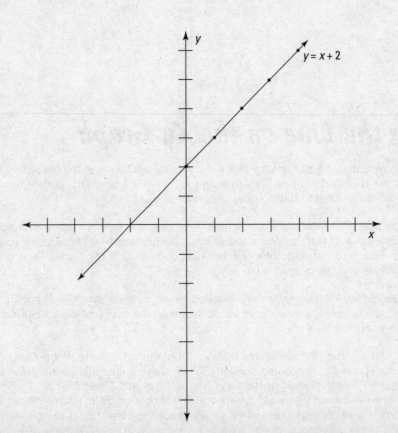

3. Graph $y = x - 1$.

a. Make an input-output table for the function $y = x - 1$ for the input values 0, 1, 2, 3, and 4.

b. Use this table to write down *xy*-coordinates for five points on this function.

c. Plot these five points to draw the function $y = x - 1$ as a line on the graph.

Solve It

4. Graph $y = 2x$.

a. Make an input-output table for the function $y = 2x$ for the input values 0, 1, 2, 3, and 4. (The notation $2x$ means $2 \times x$.)

b. Use this table to write down *xy*-coordinates for five points on this function.

c. Plot these five points to draw the function $y = 2x$ as a line on the graph.

Solve It

5. Graph $y = 3x - 5$.

a. Make an input-output table for the function $y = 3x - 5$ for the input values 0, 1, 2, 3, and 4. (The notation $3x$ means $3 \times x$.)

b. Use this table to write down *xy*-coordinates for five points on this function.

c. Plot these five points to draw the function $y = 3x - 5$ as a line on the graph.

Solve It

6. Graph $y = \frac{x}{2} + 3$.

a. Make an input-output table for the function $y = \frac{x}{2} + 3$ for the input values –2, 0, 2, and 4.

b. Use this table to write down *xy*-coordinates for four points on this function.

c. Plot these four points to draw the function $y = \frac{x}{2} + 3$ as a line on the graph.

Solve It

Answers to Problems in Getting Graphic: Xy-Graphs

The following are the answers to the practice questions presented in this chapter.

1 See the following graph.

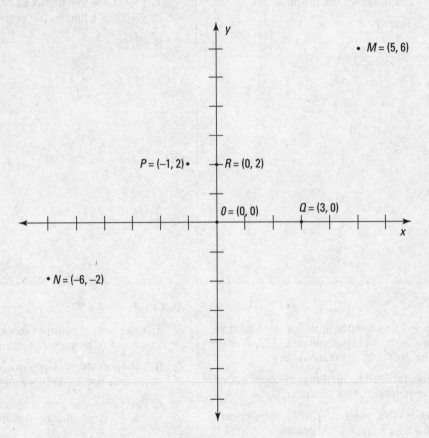

2 $S = (4,0)$, $T = (1,3)$, $U = (-5,4)$, $V = (-3,-3)$, $W = (0,-6)$, and $X = (1,-1)$.

3 Graph $y = x - 1$.

a. Here's the input-output table for the function $y = x - 1$.

Input Value x	x − 1	Output Value y
0	0 − 1	−1
1	1 − 1	0
2	2 − 1	1
3	3 − 1	2
4	4 − 1	3

b. (0,–1), (1,0), (2,1), (3,2), and (4,3).

c. See the following graph.

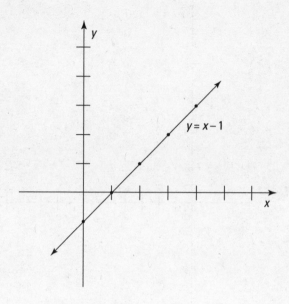

$y = x - 1$

4   Graph $y = 2x$.

a. Make the input-output table for $y = 2x$.

Input Value x	2x	Output Value y
0	2×0	0
1	2×1	2
2	2×2	4
3	2×3	6
4	2×4	8

b. (0,0), (1,2), (2,4), (3,6), and (4,8).

c. See the following graph.

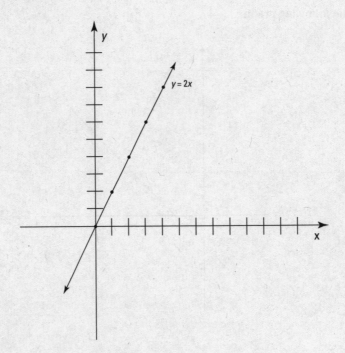

5   Graph $y = 3x - 5$.

a. Here's the input-output table for $y = 3x - 5$.

Input Value x	3x − 5	Output Value y
0	$(3 \times 0) - 5$	−5
1	$(3 \times 1) - 5$	−2
2	$(3 \times 2) - 5$	1
3	$(3 \times 3) - 5$	4
4	$(3 \times 4) - 5$	7

b. (0,–5), (1,–2), (2,1), (3,4), and (4,7).

c. See the following graph.

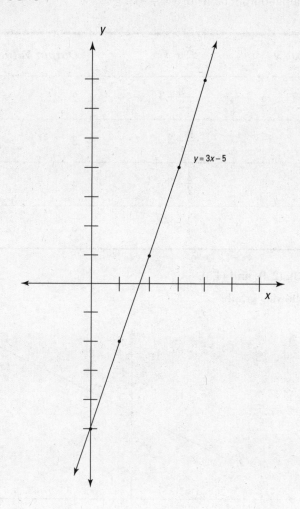

$y = 3x - 5$

6 Graph $y = \frac{x}{2} + 3$.

a. Fill in the input-output table for $y = \frac{x}{2} + 3$:

Input Value x	$\frac{x}{2} + 3$	Output Value y
–2	$-\frac{2}{2} + 3$	2
0	$\frac{0}{2} + 3$	3
2	$\frac{2}{2} + 3$	4
4	$\frac{4}{2} + 3$	5

b. (–2,2), (0,3), (2,4), and (4,5).

c. See the following graph.

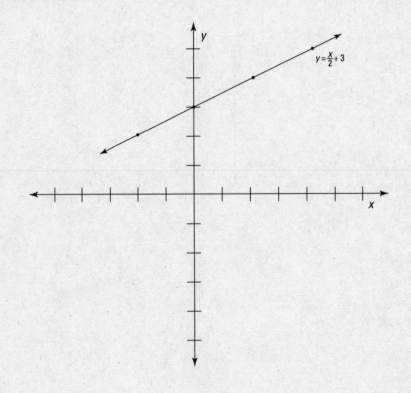

Part IV
The X Factor: Introducing Algebra

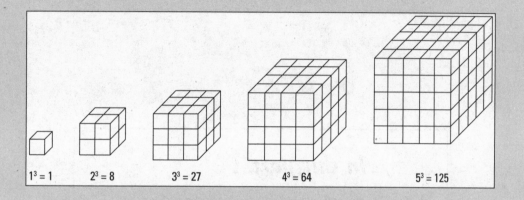

$1^3 = 1$ $2^3 = 8$ $3^3 = 27$ $4^3 = 64$ $5^3 = 125$

See how to solve systems of equations in algebra in a free article at www.dummies.com/extras/basicmathprealgebrawb.

In this part. . .

- Evaluate algebraic expressions by plugging in numbers for variables.
- Simplify expressions by removing parentheses and combining like terms.
- Factor expressions using the greatest common factor of all the terms.
- Solve algebraic equations by isolating the variable.
- Use cross-multiplication to solve equations with fractions.

Chapter 14

Expressing Yourself with Algebraic Expressions

. . . .

In This Chapter

▶ Evaluating algebraic expressions

▶ Breaking an expression into terms and identifying like terms

▶ Applying the Big Four operations to algebraic terms

▶ Simplifying and FOILing expressions

. . . .

In arithmetic, sometimes a box, a blank space, or a question mark stands for an unknown number, as in $2 + 3 = \square$, $9 - \underline{\quad} = 5$, or $? \times 6 = 18$. But in algebra, a letter such as x stands for an unknown number. A letter in algebra is called a *variable* because its value can vary from one problem to the next. For instance, in the equation $10 - x = 8$, x has a value of 2 because $10 - 2 = 8$. But in the equation $2(x) = 12$, x has a value of 6 because $2 \times 6 = 12$.

Here are a few algebra conventions you should know:

▶ **Multiplication:** When multiplying by a variable, you rarely use the multiplication operator $\times$ or $\cdot$. As in arithmetic, you can use parentheses without an operator to express multiplication. For example, $3(x)$ means $3 \times x$. Often, the multiplication operator is dropped altogether. For example, $3x$ means $3 \times x$.

▶ **Division:** The fraction bar replaces the division sign, so to express $x \div 5$, you write $\frac{x}{5}$.

▶ **Powers:** Algebra commonly uses exponents to show a variable multiplied by itself. So to express $x \times x$, you write x^2 rather than xx. To express xxx, you write x^3.

An exponent applies *only* to the variable that it follows, so in the expression $2xy^2$, only the y is squared. To make the exponent apply to both variables, you'd have to group them in parentheses, as in $2(xy)^2$.

In this chapter, you find out how to read, evaluate, break down, and simplify basic algebraic expressions. I also show you how to add, subtract, and multiply algebraic expressions.

Plug It In: Evaluating Algebraic Expressions

An arithmetic expression is any sequence of numbers and operators that makes sense when placed on one side of an equal sign. An *algebraic expression* is similar, except that it includes at least one variable (such as x). Just as with arithmetic expressions, an algebraic expression can be *evaluated* — that is, reduced to a single number. In algebra, however, this evaluation depends on the value of the variable.

Here's how to evaluate an algebraic expression when you're given a value for every variable:

1. **Substitute the value for every variable in the expression.**

2. **Evaluate the expression using the order of operations, as I show you in Chapter 4.**

Q. Evaluate the algebraic expression $x^2 - 2x + 5$ when $x = 3$.

A. **8.** First, substitute 3 for x in the expression:

$$x^2 - 2x + 5 = 3^2 - 2(3) + 5$$

Evaluate the expression using the order of operations. Start by evaluating the power:

$$= 9 - 2(3) + 5$$

Next, evaluate the multiplication:

$$= 9 - 6 + 5$$

Finally, evaluate the addition and subtraction from left to right:

$$= 3 + 5 = 8$$

Q. Evaluate the algebraic expression $3x^2 + 2xy^4 - y^3$ when $x = 5$ and $y = 2$.

A. **227.** Plug in 5 for x and 3 for y in the expression:

$$3x^2 + 2xy^4 - y^3 = 3(5^2) + 2(5)(2^4) - 2^3$$

Evaluate the expression using the order of operations. Start by evaluating the three powers:

$$= 3(25) + 2(5)(16) - 8$$

Next, evaluate the multiplication:

$$= 75 + 160 - 8$$

Finally, evaluate the addition and subtraction from left to right:

$$= 235 - 8 = 227$$

1. Evaluate the expression $x^2 + 5x + 4$ when $x = 3$.

Solve It

2. Find the value of $5x^4 + x^3 - x^2 + 10x + 8$ when $x = -2$.

Solve It

3. Evaluate the expression $x(x^2 - 6)(x - 7)$ when $x = 4$.

Solve It

4. Evaluate $\dfrac{(x-9)^4}{(x+4)^3}$ when $x = 6$.

Solve It

5. Find the value of $3x^2 + 5xy + 4y^2$ when $x = 5$ and $y = 7$.

Solve It

6. Evaluate the expression $x^6y - 5xy^2$ when $x = -1$ and $y = 9$.

Solve It

Knowing the Terms of Separation

Algebraic expressions begin to make more sense when you understand how they're put together, and the best way to understand this is to take them apart and know what each part is called. Every algebraic expression can be broken into one or more terms. A *term* is any chunk of symbols set off from the rest of the expression by either an addition or subtraction sign. For example,

- The expression $-7x + 2xy$ has two terms: $-7x$ and $2xy$.
- The expression $x^4 - \dfrac{x^2}{5} - 2x + 2$ has four terms: x^4, $\dfrac{-x^2}{5}$, $-2x$, and 2.
- The expression $8x^2y^3z^4$ has only one term: $8x^2y^3z^4$.

When you separate an algebraic expression into terms, group the plus or minus sign with the term that immediately follows it. Then you can rearrange terms in any order you like. After the terms are separated, you can drop the plus signs.

When a term doesn't have a variable, it's called a *constant*. (*Constant* is just a fancy word for *number* when you're talking about terms in an expression.) When a term has a variable, it's called an *algebraic term*. Every algebraic term has two parts:

- The *coefficient* is the signed numerical part of a term — that is, the number and the sign (+ or –) that go with that term. Typically, the coefficients 1 and –1 are dropped from a term, so when a term appears to have no coefficient, its coefficient is either 1 or –1, depending on the sign of that term. The coefficient of a constant is the constant itself.

- The *variable part* is everything else other than the coefficient.

When two terms have the same variable part, they're called *like terms* or *similar terms*. For two terms to be like, both the letters and their exponents have to be exact matches. For example, the terms $2x$, $-12x$, and $834x$ are all like terms because their variable parts are all x. Similarly, the terms $17xy$, $1,000xy$, and $0.625xy$ are all like terms because their variable parts are all xy.

Q. In the expression $3x^2 - 2x - 9$, identify which terms are algebraic and which are constants.

A. **$3x^2$ and $-2x$ are both algebraic terms, and -9 is a constant.**

Q. Identify the coefficient of every term in the expression $2x^4 - 5x^3 + x^2 - x - 9$.

A. **2, –5, 1, –1, and –9.** The expression has five terms: $2x^4$, $-5x^3$, x^2, $-x$, -9. The coefficient of $2x^4$ is 2, and the coefficient of $-5x^3$ is –5. The term x^2 appears to have no coefficient, so its coefficient is 1. The term $-x$ also appears to have no coefficient, so its coefficient it –1. The term –9 is a constant, so its coefficient is –9.

7. Write down all the terms in the expression $7x^2yz - 10xy^2z + 4xyz^2 + y - z + 2$. Which are algebraic terms, and which are constants?

Solve It

8. Identify the coefficient of every term in $-2x^5 + 6x^4 + x^3 - x^2 - 8x + 17$.

Solve It

9. Name every coefficient in the expression $-x^3y^3z^3 - x^2y^2z^2 + xyz - x$.

Solve It

10. In the expression $12x^3 + 7x^2 - 2x - x^2 - 8x^4 + 99x + 99$, identify any sets of like terms.

Solve It

Adding and Subtracting Like Terms

You can add and subtract only like algebraic terms (see the preceding section). In other words, the variable parts need to match. To add two like terms, simply add the coefficients and keep the variable parts of the terms the same. Subtraction works much the same: Find the difference between their coefficients and keep the same variable part.

Q. What is $3x + 5x$?

A. **8x.** The variable part of both terms is x, so you can add:

$$3x + 5x = (3 + 5)x = 8x$$

Q. What is $24x^3 - 7x^3$?

A. **17x^3.** Subtract the coefficients:

$$24x^3 - 7x^3 = (24 - 7)x^3 = 17x^3$$

11. Add $4x^2y + (-9x^2y)$.

Solve It

12. Add $x^3y^2 + 20x^3y^2$.

Solve It

13. Subtract $-2xy^4 - 20xy^4$.

Solve It

14. Subtract $-xyz - (-xyz)$.

Solve It

Multiplying and Dividing Terms

Unlike addition and subtraction, you can multiply *any* two algebraic terms, whether they're like or nonlike. Multiply two terms by multiplying their coefficients and collecting all the variables in each term into a single term. (When you collect the variables, you're simply using exponents to give a count of how many x's, y's, and so on appeared in the original terms.)

A fast way to multiply variables with exponents is to add the exponents of identical variables together.

You can also divide any two algebraic terms. Division in algebra is usually represented as a fraction. Dividing is similar to reducing a fraction to lowest terms:

1. **Reduce the coefficients to lowest terms as you would any other fraction (see Chapter 6).**

2. **Cancel out variables that appear in both the numerator and the denominator.**

A fast way to divide is to subtract the exponents of identical variables. For each variable, subtract the exponent in the numerator minus the exponent in the denominator. When a resulting exponent is a negative number, move that variable to the denominator and remove the minus sign.

Q. What is $2x(6y)$?

A. **$12xy$.** To get the final coefficient, multiply the coefficients $2 \times 6 = 12$. To get the variable part of the answer, combine the variables x and y:

$$2x(6y) = 2(6)xy = 12xy$$

Q. What is $4x(-9x)$?

A. **$-36x^2$.** To get the final coefficient, multiply the coefficients $4 \times -9 = -36$. To get the variable part of the answer, collect the variables into a single term:

$$4x(-9x) = 4(-9)xx$$

Remember that x^2 is shorthand for xx, so rewrite the answer as follows:

$$-36x^2$$

Q. What is $2x(4xy)(5y^2)$?

A. **$40x^2y^3$.** Multiply all three coefficients together and gather up the variables:

$$2x(4xy)(5y^2) = 2(4)(5)xxyy^2 = 40x^2y^3$$

In this answer, the exponent 2 that's associated with x is just a count of how many x's appear in the problem; the same is true of the exponent 3 that's associated with y.

Q. What is $(x^2y^3)(xy^5)(x^4)$?

A. **x^7y^8.** Add exponents of the three x's $(2 + 1 + 4 = 7)$ to get the exponent of x in the answer (remember that x means x^1). Add the two y exponents $(3 + 5 = 8)$ to get the exponent of y in the answer.

Q. What is $\dfrac{6x^3y^2}{3xy^2}$?

A. $2x^2$. Exponents mean repeated multiplication, so

$$\frac{6x^3y^2}{3xy^2} = \frac{6xxxyy}{3xyy}$$

Both the numerator and the denominator are divisible by 2, so reduce the coefficients just as you'd reduce a fraction:

$$= \frac{2xxxyy}{xyy}$$

Now cancel out any variables repeated in both the numerator and denominator — that is, one x and two y's both above and below the fraction bar:

$$= \frac{2xx\cancel{x}\cancel{y}\cancel{y}}{\cancel{x}\cancel{y}\cancel{y}} = 2xx$$

Rewrite this result using an exponent.

$$2x^2$$

---

15. Multiply $4x(7x^2)$.

Solve It

16. Multiply $- xy^3z^4(10x^2y^2z)(-2xz)$.

Solve It

---

17. Divide $\dfrac{6x^4y^5}{8x^4y^4}$.

Solve It

18. Divide $\dfrac{7x^2y}{21xy^3}$.

Solve It

Simplifying Expressions by Combining Like Terms

Although an algebraic expression may have any number of terms, you can sometimes make it smaller and easier to work with. This process is called *simplifying* an expression. To simplify, you *combine like terms,* which means you add or subtract any terms that have identical variable parts. (To understand how to identify like algebraic terms, see "Knowing the Terms of Separation," earlier in this chapter. "Adding and Subtracting Like Terms" shows you how to do the math.)

In some expressions, like terms may not all be next to each other. In this case, you may want to rearrange the terms to place all like terms together before combining them.

When rearranging terms, you *must* keep the sign (+ or –) with the term that follows it.

To finish, you usually rearrange the answer alphabetically, from highest exponent to lowest, and place any constant last. In other words, if your answer is $-5 + 4y^3 + x^2 + 2x - 3xy$, you'd rearrange it to read $x^2 + 2x - 3xy + 4y^3 - 5$. (This step doesn't change the answer, but it kind of cleans it up, so teachers love it.)

Q. Simplify the expression $x^2 + 2x - 7x + 1$.

A. $x^2 - 5x + 1$. The terms $2x$ and $-7x$ are like, so you can combine them:

$x^2 + \underline{2x - 7x} + 1 = x^2 + (2 - 7)x + 1$
$= x^2 - 5x + 1$

Q. Simplify the expression $4x^2 - 3x + 2 + x - 7x^2$.

A. $-3x^2 - 2x + 2$. I've begun by underlining the five terms of the expression:

$\underline{4x^2} \ \underline{- 3x} \ \underline{+ 2} \ \underline{+ x} \ \underline{- 7x^2}$

Rearrange these terms as needed so that all like terms are together:

$\underline{4x^2 - 7x^2} \ \underline{- 3x + x} + 2$

At this point, combine the two underlined pairs of like terms:

$(4 - 7)x^2 + (-3 + 1)x + 2 = -3x^2 - 2x + 2$

19. Simplify the expression $3x^2 + 5x^2 + 2x - 8x - 1$.

Solve It

20. Simplify the expression $6x^3 - x^2 + 2 - 5x^2 - 1 + x$.

Solve It

21. Simplify the expression $2x^4 - 2x^3 + 2x^2 - x + 9 + x + 7x^2$.

Solve It

22. Simplify the expression $x^5 - x^3 + xy - 5x^3 - 1 + x^3 - xy + x$.

Solve It

Simplifying Expressions with Parentheses

When an expression contains parentheses, you need to get rid of the parentheses before you can simplify the expression. Here are the four possible cases:

- **Parentheses preceded by a plus sign (+):** Just remove the parentheses. After that, you may be able to simplify the expression further by combining like terms, as I show you in the preceding section.

- **Parentheses preceded by a minus sign (–):** Change every term inside the parentheses to the opposite sign; then remove the parentheses. After the parentheses are gone, combine like terms.

- **Parentheses preceded by no sign (a term directly next to a set of parentheses):** Multiply the term next to the parentheses by every term inside the parentheses (make sure you include the plus or minus signs in your terms); then drop the parentheses. Simplify by combining like terms.

  To multiply identical variables, simply add the exponents. For instance, $x(x^2) = x^{1+2} = x^3$. Likewise, $-x^3(x^4) = -(x^{3+4}) = -x^7$.

- **Two sets of adjacent parentheses:** I discuss this case in the next section.

Q. Simplify the expression $7x + (x^2 - 6x + 4) - 5$.

A. $x^2 + x - 1$. Because a plus sign precedes this set of parentheses, you can drop the parentheses.

$$7x + (x^2 - 6x + 4) - 5$$

$$= 7x + x^2 - 6x + 4 - 5$$

Now combine like terms. I do this in two steps:

$$= \underline{7x - 6x} + x^2 + \underline{4 - 5}$$

Finally, rearrange the answer from highest exponent to lowest:

$$= x + x^2 - 1$$

Q. Simplify the expression $x - 3x(x^3 - 4x^2 + 2) + 8x^4$.

A. $5x^4 + 12x^3 - 5x$. The term $-3x$ precedes this set of parentheses without a sign in between, so multiply every term inside the parentheses by $-3x$ and then drop the parentheses:

$$x - 3x(x^3 - 4x^2 + 2) + 8x^4$$

$$= x + -3x(x^3) + -3x(-4x^2) + -3x(2) + 8x^4$$

$$= x - 3x^4 + 12x^3 - 6x + 8x^4$$

Now combine like terms. I do this in two steps:

$$= \underline{x - 6x} - \underline{3x^4 + 8x^4} + 12x^3$$

$$= -5x + 5x^4 + 12x^3$$

As a final step, arrange the terms in order from highest to lowest exponent:

$$= 5x^4 + 12x^3 - 5x$$

23. Simplify the expression $3x^3 + (12x^3 - 6x) + (5 - x)$.

Solve It

24. Simplify the expression $2x^4 - (-9x^2 + x) + (x + 10)$.

Solve It

25. Simplify the expression $x - (x^3 - x - 5) + 3(x^2 - x)$.

Solve It

26. Simplify the expression $-x^3(x^2 + x) - (x^5 - x^4)$.

Solve It

FOILing: Dealing with Two Sets of Parentheses

When an expression has two sets of parentheses next to each other, you need to multiply every term inside the first set of parentheses by every term in the second set. This process is called FOILing. The word *FOIL* is a memory device for the words *First, Outside, Inside, Last,* which helps keep track of the multiplication when both sets of parentheses have two terms each.

When two sets of adjacent parentheses are part of a larger expression, FOIL the contents of the parentheses and place the results into one set of parentheses. Then remove this set of parentheses using one of the rules I show you in the preceding section.

Q. Simplify the expression $(x + 4)(x - 3)$.

A. $x^2 + x - 12$. Start by multiplying the two first terms:

$$(\underline{x} + 4)(\underline{x} - 3) \qquad x \times x = x^2$$

Next, multiply the two outside terms:

$$(\underline{x} + 4)(x \underline{- 3}) \qquad x \times -3 = -3x$$

Now multiply the two inside terms:

$$(x + \underline{4})(\underline{x} - 3) \qquad 4 \times x = 4x$$

Finally, multiply the two last terms:

$$(x + \underline{4})(x \underline{- 3}) \qquad 4 \times -3 = -12$$

Add all four results together and simplify by combining like terms:

$$(x + 4)(x - 3) = x^2 \underline{- 3x + 4x} - 12 = x^2 + x - 12$$

Q. Simplify the expression $x^2 - (-2x + 5)(3x - 1) + 9$.

A. $7x^2 - 17x + 14$. Begin by FOILing the parentheses. Start by multiplying the two first terms:

$$(\underline{-2x} + 5)(\underline{3x} - 1) \qquad -2x \times 3x = -6x^2$$

Multiply the two outside terms:

$$(\underline{-2x} + 5)(3x \underline{- 1}) \qquad -2x \times -1 = 2x$$

Multiply the two inside terms:

$$(-2x \underline{+ 5})(\underline{3x} - 1) \qquad 5 \times 3x = 15x$$

Finally, multiply the two last terms:

$$(-2x + \underline{5})(3x \underline{- 1}) \qquad 5 \times -1 = -5$$

Add these four products together and put the result inside one set of parentheses, replacing the two sets of parentheses that were originally there:

$$x^2 - (-2x + 5)(3x - 1) + 9$$

$$= x^2 - (-6x^2 + 2x + 15x - 5) + 9$$

The remaining set of parentheses is preceded by a minus sign, so change the sign of every term in there to its opposite and drop the parentheses:

$$= x^2 + 6x^2 - 2x - 15x + 5 + 9$$

At this point, you can simplify the expression by combining like terms:

$$= \underline{x^2 + 6x^2} \underline{- 2x - 15x} + \underline{5 + 9}$$

$$= 7x^2 - 17x + 14$$

27. Simplify the expression $(x + 7)(x - 2)$.

Solve It

28. Simplify the expression $(x - 1)(-x - 9)$.

Solve It

29. Simplify the expression $6x - (x - 2)(x - 4) + 7x^2$.

Solve It

30. Simplify the expression $3 - 4x(x^2 + 1)(x - 5) + 2x^3$.

Solve It

Answers to Problems in Expressing Yourself with Algebraic Expressions

The following are the answers to the practice questions presented in this chapter.

1 $x^2 + 5x + 4 =$ **28.** Substitute 3 for x in the expression:

$$x^2 + 5x + 4 = 3^2 + 5(3) + 4$$

Evaluate the expression using the order of operations. Start with the power:

$$= 9 + 5(3) + 4$$

Continue by evaluating the multiplication:

$$= 9 + 15 + 4$$

Finish up by evaluating the addition from left to right:

$$= 24 + 4 = 28$$

2 $5x^4 + x^3 - x^2 + 10x + 8 =$ **56.** Plug in –2 for every x in the expression:

$$5x^4 + x^3 - x^2 + 10x + 8 = 5(-2)^4 + (-2)^3 - (-2)^2 + 10(-2) + 8$$

Evaluate the expression using the order of operations. Start with the powers:

$$= 5(16) + -8 - 4 + 10(-2) + 8$$

Do the multiplication:

$$= 80 + -8 - 4 + -20 + 8$$

Finish up by evaluating the addition and subtraction from left to right:

$$= 72 - 4 + -20 + 8 = 68 + -20 + 8 = 48 + 8 = 56$$

3 $x(x^2 - 6)(x - 7) =$ **–120.** Substitute 4 for x in the expression:

$$x(x^2 - 6)(x - 7) = 4(4^2 - 6)(4 - 7)$$

Follow the order of operations as you evaluate the expression. Starting inside the first set of parentheses, evaluate the power and then the subtraction:

$$= 4(16 - 6)(4 - 7) = 4(10)(4 - 7)$$

Find the contents of the remaining set of parentheses:

$$= 4(10)(-3)$$

Evaluate the multiplication from left to right:

$$= 40(-3) = -120$$

4 $\dfrac{(x-9)^4}{(x+4)^3} = \dfrac{81}{1,000}$ **(or 0.081).** Replace every x in the expression with a 6:

$$\frac{(x-9)^4}{(x+4)^3} = \frac{(6-9)^4}{(6+4)^3}$$

Follow the order of operations. Evaluate the contents of the set of parentheses in the numerator and then in the denominator:

$$= \frac{(-3)^4}{(6+4)^3} = \frac{(-3)^4}{10^3}$$

Continue by evaluating the powers from top to bottom:

$$= \frac{81}{10^3} = \frac{81}{1,000}$$

You can also express this fraction as the decimal 0.081.

5 $3x^2 + 5xy + 4y^2 =$ **446.** Substitute 5 for x and 7 for y in the expression:

$$3x^2 + 5xy + 4y^2 = 3(5)^2 + 5(5)(7) + 4(7)^2$$

Evaluate using the order of operations. Start with the two powers:

$$= 3(25) + 5(5)(7) + 4(49)$$

Evaluate the multiplication from left to right:

$$= 75 + 175 + 196$$

Finally, do the addition from left to right:

$$= 250 + 196 = 446$$

6 $x^6y - 5xy^2 =$ **414.** Plug in –1 for each x and 9 for each y in the expression:

$$x^6y - 5xy^2 = (-1)^6(9) - 5(-1)(9)^2$$

Follow the order of operations. Start by evaluating the two powers:

$$= 1(9) - 5(-1)(81)$$

Continue by evaluating the multiplication from left to right:

$$= 9 - (-5)(81) = 9 - (-405)$$

Finally, do the subtraction:

$$= 414$$

7   In $7x^2yz - 10xy^2z + 4xyz^2 + y - z + 2$, **the algebraic terms are $7x^2yz$, $-10xy^2z$, $4xyz^2$, y, and $-z$; the constant is 2.**

8   In $-2x^5 + 6x^4 + x^3 - x^2 - 8x + 17$, **the six coefficients, in order, are –2, 6, 1, –1, –8, and 17.**

9   In $-x^3y^3z^3 - x^2y^2z^2 + xyz - x$, **the four coefficients, in order, are –1, –1, 1, and –1.**

10   In $12x^3 + 7x^2 - 2x - x^2 - 8x^4 + 99x + 99$, **$7x^2$ and $-x^2$** are like terms (the variable part is x^2); **$-2x$ and $99x$** are also like terms (the variable part is x).

11 $4x^2y + -9x^2y = (4 + -9)x^2y =$ **$-5x^2y$.**

12 $x^3y^2 + 20x^3y^2 = (1 + 20)x^3y^2 =$ **$21x^3y^2$.**

13 $-2xy^4 - 20xy^4 = (-2 - 20)xy^4 =$ **$-22xy^4$.**

14 $-xyz - (-xyz) = [-1 - (-1)]xyz = (-1 + 1)xyz =$ **0.**

15 $4x(7x^2) =$ **$28x^3$.** Multiply the two coefficients to get the coefficient of the answer; then gather the variables into one term:

$$4x(7x^2) = 4(7)xx^2 = 28x^3$$

16 $-xy^3z^4(10x^2y^2z)(-2xz) = \mathbf{20x^4y^5z^6}$. Multiply the coefficients ($-1 \times 10 \times -2 = 20$) to get the coefficient of the answer. Add the x exponents ($1 + 2 + 1 = 4$) to get the exponent of x in the answer. Add the y exponents ($3 + 2 = 5$) to get the exponent of y in the answer. And add the z exponents ($4 + 1 + 1 = 6$) to get the exponent of z in the answer.

17 $\dfrac{6x^4y^5}{8x^4y^4} = \dfrac{\mathbf{3y}}{\mathbf{4}}$. Reduce the coefficients of the numerator and denominator just as you'd reduce a fraction:

$$\frac{6x^4y^5}{8x^4y^4} = \frac{3x^4y^5}{4x^4y^4}$$

To get the x exponent of the answer, take the x exponent in the numerator minus the x exponent in the denominator: $4 - 4 = 0$, so the x's cancel:

$$= \frac{3y^5}{4y^4}$$

To get the y exponent of the answer, take the y exponent in the numerator minus the y exponent in the denominator: $5 - 4 = 1$, so you have only y^1, or y, in the numerator:

$$= \frac{3y}{4}$$

18 $\dfrac{7x^2y}{21xy^3} = \dfrac{\mathbf{x}}{\mathbf{3y^2}}$. Reduce the coefficients of the numerator and denominator just as you'd reduce a fraction:

$$\frac{7x^2y}{21xy^3} = \frac{x^2y}{3xy^3}$$

Take the x exponent in the numerator minus the x exponent in the denominator ($2 - 1 = 1$) to get the x exponent of the answer:

$$= \frac{xy}{3y^3}$$

To get the y exponent, take the y exponent in the numerator minus the y exponent in the denominator ($1 - 3 = -2$):

$$= \frac{xy^{-2}}{3}$$

To finish up, remove the minus sign from the y exponent and move this variable to the denominator:

$$= \frac{x}{3y^2}$$

19 $3x^2 + 5x^2 + 2x - 8x - 1 = \mathbf{8x^2 - 6x - 1}$. Combine the following underlined like terms:

$$\underline{3x^2 + 5x^2} + \underline{2x - 8x} - 1$$
$$= 8x^2 - 6x - 1$$

20 $6x^3 - x^2 + 2 - 5x^2 - 1 + x = \mathbf{6x^3 - 6x^2 + x + 1}$. Rearrange the terms so like terms are next to each other:

$$6x^3 - x^2 + 2 - 5x^2 - 1 + x$$

$$= 6x^3 \underline{- x^2 - 5x^2} + x + \underline{2 - 1}$$

Now combine the underlined like terms:

$$= 6x^3 - 6x^2 + x + 1$$

21 $2x^4 - 2x^3 + 2x^2 - x + 9 + x + 7x^2 = \mathbf{2x^4 - 2x^3 + 9x^2 + 9}$. Put like terms next to each other:

$$2x^4 - 2x^3 + 2x^2 - x + 9 + x + 7x^2$$

$$= 2x^4 - 2x^3 + \underline{2x^2 + 7x^2} \underline{- x + x} + 9$$

Now combine the underlined like terms:

$$= 2x^4 - 2x^3 + 9x^2 + 9$$

Note that the two x terms cancel each other out.

22 $x^5 - x^3 + xy - 5x^3 - 1 + x^3 - xy + x = \mathbf{x^5 - 5x^3 + x - 1}$. Rearrange the terms so like terms are next to each other:

$$x^5 - x^3 + xy - 5x^3 - 1 + x^3 - xy + x$$

$$= x^5 \underline{- x^3 - 5x^3 + x^3} + \underline{xy - xy} + x - 1$$

Now combine the underlined like terms:

$$= x^5 - 5x^3 + x - 1$$

Note that the two xy terms cancel each other out.

23 $3x^3 + (12x^3 - 6x) + (5 - x) = \mathbf{15x^3 - 7x + 5}$. A plus sign precedes both sets of parentheses, so you can drop both sets:

$$3x^3 + (12x^3 - 6x) + (5 - x)$$

$$= 3x^3 + 12x^3 - 6x + 5 - x$$

Now combine like terms:

$$= \underline{3x^3 + 12x^3} \underline{- 6x - x} + 5$$

$$= 15x^3 - 7x + 5$$

24 $2x^4 - (-9x^2 + x) + (x + 10) = \mathbf{2x^4 + 9x^2 + 10}$. A minus sign precedes the first set of parentheses, so change the sign of every term inside this set and then drop these parentheses:

$$2x^4 - (-9x^2 + x) + (x + 10)$$

$$= 2x^4 + 9x^2 - x + (x + 10)$$

A plus sign precedes the second set of parentheses, so just drop this set:

$$= 2x^4 + 9x^2 - x + x + 10$$

Now combine like terms:

$$= 2x^4 + 9x^2 + 10$$

25 $x - (x^3 - x - 5) + 3(x^2 - x) = -x^3 + 3x^2 - x + 5$. A minus sign precedes the first set of parentheses, so change the sign of every term inside this set and then drop these parentheses:

$$x - (x^3 - x - 5) + 3(x^2 - x)$$
$$= x - x^3 + x + 5 + 3(x^2 - x)$$

You have no sign between the term 3 and the second set of parentheses, so multiply every term inside these parentheses by 3 and then drop the parentheses:

$$= x - x^3 + x + 5 + 3x^2 - 3x$$

Now combine like terms:

$$= \underline{x + x - 3x} - x^3 + 5 + 3x^2$$
$$= -x - x^3 + 5 + 3x^2$$

Rearrange the answer so that the exponents are in descending order:

$$= -x^3 + 3x^2 - x + 5$$

26 $-x^3(x^2 + x) - (x^5 - x^4) = -2x^5$. You have no sign between the term $-x^3$ and the first set of parentheses, so multiply $-x^3$ by every term inside this set and then drop these parentheses:

$$-x^3(x^2 + x) - (x^5 - x^4)$$
$$= -x^5 - x^4 - (x^5 - x^4)$$

A minus sign precedes the second set of parentheses, so change every term inside this set and then drop the parentheses:

$$= -x^5 - x^4 - x^5 + x^4$$

Combine like terms, noting that the x^4 terms cancel out:

$$= \underline{-x^5 - x^5} - x^4 + x^4$$
$$= -2x^5$$

27 $(x + 7)(x - 2) = x^2 + 5x - 14$. Begin by FOILing the parentheses. Start by multiplying the two first terms:

$$(\underline{x} + 7)(\underline{x} - 2) \qquad x \times x = x^2$$

Multiply the two outside terms:

$$(\underline{x} + 7)(x \underline{- 2}) \qquad x \times -2 = -2x$$

Multiply the two inside terms:

$$(x + \underline{7})(\underline{x} - 2) \qquad 7 \times x = 7x$$

Finally, multiply the two last terms:

$$(x + \underline{7})(x \underline{- 2}) \qquad 7 \times -2 = -14$$

Add these four products together and simplify by combining like terms:

$$x^2 \underline{- 2x + 7x} - 14 = x^2 + 5x - 14$$

28 $(x - 1)(-x - 9) = -x^2 - 8x + 9$. FOIL the parentheses. Multiply the two first terms, the two outside terms, the two inside terms, and the two last terms:

$(\underline{x} - 1)(\underline{-x} - 9)$ $x \times -x = -x^2$

$(\underline{x} - 1)(-x \underline{- 9})$ $x \times -9 = -9x$

$(x \underline{- 1})(\underline{-x} - 9)$ $-1 \times -x = x$

$(x \underline{- 1})(-x \underline{- 9})$ $-1 \times -9 = 9$

Add these four products together and simplify by combining like terms:

$-x^2 \underline{- 9x + x} + 9 = -x^2 - 8x + 9$

29 $6x - (x - 2)(x - 4) + 7x^2 = \mathbf{6x^2 + 12x - 8}$. Begin by FOILing the parentheses: Multiply the first, outside, inside, and last terms:

$(\underline{x} - 2)(\underline{x} - 4)$ $x \times = x^2$

$(\underline{x} - 2)(x \underline{- 4})$ $x \times -4 = -4x$

$(x \underline{- 2})(\underline{x} - 4)$ $-2 \times x = -2x$

$(x \underline{- 2})(x \underline{- 4})$ $-2 \times -4 = 8$

Add these four products together and put the result inside one set of parentheses, replacing the two sets of parentheses that were originally there:

$6x - (x - 2)(x - 4) + 7x^2 = 6x - (x^2 - 4x - 2x + 8) + 7x^2$

The remaining set of parentheses is preceded by a minus sign, so change the sign of every term in there to its opposite and drop the parentheses:

$= 6x - x^2 + 4x + 2x - 8 + 7x^2$

Now simplify the expression by combining like terms and reordering your solution:

$= \underline{6x + 4x + 2x} \underline{- x^2 + 7x^2} - 8$

$= 6x^2 + 12x - 8$

30 $3 - 4x(x^2 + 1)(x - 5) + 2x^3 = \mathbf{-4x^4 + 22x^3 - 4x^2 + 20x + 3}$. Begin by FOILing the parentheses, multiplying the first, outside, inside, and last terms:

$(\underline{x^2} + 1)(\underline{x} - 5)$ $x^2 \times x = x^3$

$(\underline{x^2} + 1)(x \underline{- 5})$ $x^2 \times -5 = -5x^2$

$(x^2 + \underline{1})(\underline{x} - 5)$ $1 \times x = x$

$(x^2 + \underline{1})(x \underline{- 5})$ $1 \times -5 = -5$

Add these four products together and put the result inside one set of parentheses, replacing the two sets of parentheses that were originally there:

$3 - 4x(x^2 + 1)(x - 5) + 2x^3 = 3 - 4x(x^3 - 5x^2 + x - 5) + 2x^3$

The remaining set of parentheses is preceded by the term $-4x$ with no sign in between, so multiply $-4x$ by every term in there and then drop the parentheses:

$= 3 - 4x^4 + 20x^3 - 4x^2 + 20x + 2x^3$

Now simplify the expression by combining like terms:

$= -4x^4 + \underline{20x^3 + 2x^3} - 4x^2 + 20x + 3$

$= -4x^4 + 22x^3 - 4x^2 + 20x + 3$

Chapter 15

Finding the Right Balance: Solving Algebraic Equations

....

In This Chapter

▶ Solving simple algebraic equations without algebra

▶ Understanding the balance scale method

▶ Moving terms across the equal sign

▶ Using cross-multiplication to simplify equations

....

In this chapter, you use your skills in doing the Big Four operations and simplifying algebraic expressions (see Chapter 14) to solve algebraic equations — that is, equations with one or more variables (such as *x*). Solving an equation means you figure out the value of the variable. First, I show you how to solve very simple equations for *x* without using algebra. Then, as the problems get tougher, I show you a variety of methods to figure out the value of *x*.

Solving Simple Algebraic Equations

You don't always need algebra to solve an algebraic equation. Here are three ways to solve simpler problems:

✔ **Inspection:** For the very simplest algebra problems, *inspection* — just looking at the problem — is enough. The answer just jumps out at you.

✔ **Rewriting the problem:** In slightly harder problems, you may be able to rewrite the problem so you can find the answer. In some cases, this involves using inverse operations; in other cases, you can use the same operation with the numbers switched around. (I introduce inverse operations in Chapter 2.)

✔ **Guessing and checking:** When problems are just a wee bit tougher, you can try guessing the answer and then checking to see whether you're right. Check by substituting your guess for *x*.

When your guess is wrong, you can usually tell whether it's too high or too low. Use this information to guide your next guess.

In some cases, you can simplify the problem before you begin solving. On either side of the equal sign, you can rearrange the terms and combine like terms as I show you in Chapter 14. After the equation is simplified, use any method you like to find *x*.

Q. In the equation $x + 3 = 10$, what's the value of x?

A. $x = 7$. You can solve this problem through simple inspection. Because $7 + 3 = 10$, $x = 7$.

Q. Find the value of x in the equation $8x - 20 = 108$.

A. $x = 16$. Guess what you think the answer may be. For example, perhaps it's $x = 10$:

$$8(10) - 20 = 80 - 20 = 60$$

Because 60 is less than 108, that guess was too low. Try a higher number — say, $x = 20$:

$$8(20) - 20 = 160 - 20 = 140$$

Because 140 is greater than 108, this guess was too high. But now you know that the right answer is between 10 and 20. So now try $x = 15$:

$$8(15) - 20 = 120 - 20 = 100$$

The number 100 is only a little less than 108, so this guess was only a little too low. Try $x = 16$:

$$8(16) - 20 = 128 - 20 = 108$$

The result 108 is correct, so $x = 16$.

Q. Solve the equation $7x = 224$ for x.

A. $x = 32$. Turn the problem around using the inverse of multiplication, which is division:

$$7 \times x = 224 \text{ means } 224 \div 7 = x$$

Now you can solve the problem easily using long division (I don't show this step, but for practice with long division, see Chapter 1):

$$224 \div 7 = 32$$

Q. Solve for x: $8x^2 - x + x^2 + 4x - 9x^2 = 18$.

A. $x = 6$. Rearrange the expression on the left side of the equation so that all like terms are next to each other:

$$8x^2 + x^2 - 9x^2 - x + 4x = 18$$

Combine like terms:

$$3x = 18$$

Notice that the x^2 terms cancel each other out. Because $3(6) = 18$, you know that $x = 6$.

1. Solve for x in each case just by looking at the equation.

   a. $x + 5 = 13$

   b. $18 - x = 12$

   c. $4x = 44$

   d. $\dfrac{30}{x} = 3$

Solve It

2. Use the correct inverse operation to rewrite and solve each problem.

   a. $x + 41 = 97$

   b. $100 - x = 58$

   c. $13x = 273$

   d. $\dfrac{238}{x} = 17$

Solve It

3. Find the value of x in each equation by guessing and checking.

   a. $19x + 22 = 136$

   b. $12x - 17 = 151$

   c. $19x - 8 = 600$

   d. $x^2 + 3 = 292$

Solve It

4. Simplify the equation and then solve for x using any method you like:

   a. $x^5 - 16 + x + 20 - x^5 = 24$

   b. $5xy + x - 2xy + 27 - 3xy = 73$

   c. $6x - 3 + x^2 - x + 8 - 5x = 30$

   d. $-3 + x^3 + 4 + x - x^3 - 1 = 2xy + 7 - x - 2xy + x$

Solve It

Equality for All: Using the Balance Scale to Isolate x

Think of an equation as a balance scale, one of the classic ones that has a horizontal beam with little weighing pans hanging from each end. The equal sign means that both sides hold the same amount and, therefore, balance each other out. To keep that equal sign, you have to maintain that balance. Therefore, whatever you do to one side of the equation, you have to do to the other.

For example, the equation $4 + 2 = 6$ is in balance because both sides are equal. If you want to add 1 to one side of the equation, you need to add 1 to the other side to keep the balance. Notice that the equation stays balanced, because each side equals 7:

$$4 + 2 + 1 = 6 + 1$$

You can apply any of the Big Four operations to the equation, provided that you keep the equation balanced at all times. For example, here's how you multiply both sides of the original equation by 10. Note that the equation stays balanced, because each side equals 60:

$$10(4 + 2) = 10(6)$$

The simple concept of the balance scale is the heart and soul of algebra. When you understand how to keep the scale in balance, you can solve algebraic equations by *isolating x* — that is, by getting x alone on one side of the equation and everything else on the other side. For most basic equations, isolating x is a three-step process:

1. **Add or subtract the same number from each side to get all constants (non-x terms) on one side of the equation.**

   On the other side of the equation, the constants should cancel each other out and equal 0.

2. **Add or subtract to get all x terms on the other side of the equation.**

   The x term that's still on the same side as the constant should cancel out.

3. **Divide to isolate x.**

Q. Use the balance scale method to find the value of x in the equation $5x - 6 = 3x + 8$.

A. $x = 7$. To get all constants on the right side of the equation, add 6 to both sides, which causes the -6 to cancel out on the left side of the equation:

$$
\begin{array}{r}
5x - 6 = 3x + 8 \\
+6 + 6 \\
\hline
5x = 3x + 14
\end{array}
$$

The right side of the equation still contains a $3x$. To get all x terms on the left

side of the equation, subtract $3x$ from both sides:

$$
\begin{array}{r}
5x = 3x + 14 \\
-3x -3x \\
\hline
2x = 14
\end{array}
$$

Divide by 2 to isolate x:

$$\frac{2x}{2} = \frac{14}{2}$$

$$x = 7$$

5. Use the balance scale method to find the value of x in the equation $9x - 2 = 6x + 7$.

Solve It

6. Solve the equation $10x - 10 = 8x + 12$ using the balance scale method.

Solve It

7. Find the value of x in $4x - 17 = x + 22$.

Solve It

8. Solve for x: $15x - 40 = 11x + 4$.

Solve It

Switching Sides: Rearranging Equations to Isolate x

When you understand how to keep equations in balance (as I show you in the preceding section), you can use a quicker method to solve algebra problems. The shortcut is to *rearrange the equation* by placing all x terms on one side of the equal sign and all constants (non-x terms) on the other side. Essentially, you're doing the addition and subtraction without showing it. You can then isolate x.

Like the balance scale method, solving for x by rearranging the equation is a three-step process; however, the steps usually take less time to write:

1. **Rearrange the terms of the equation so that all x terms are on one side of the equation and all constants (non-x terms) are on the other side.**

   When you move a term from one side of the equal sign to the other, always negate that term. That is, if the term is positive, make it negative; if the term is negative, make it positive.

2. **Combine like terms on both sides of the equation.**

3. **Divide to isolate x.**

When one or both sides of the equation contain parentheses, remove them (as I show you in Chapter 14). Then use these three steps to solve for x.

Q. Find the value of x in the equation $7x - 6 = 4x + 9$.

A. $x = 5$. Rearrange the terms of the equation so that the x terms are on one side and the constants are on the other. I do this in two steps:

$$7x - 6 = 4x + 9$$

$$7x = 4x + 9 + 6$$

$$7x - 4x = 9 + 6$$

Combine like terms on both sides of the equation:

$$3x = 15$$

Divide by 3 to isolate x:

$$\frac{3x}{3} = \frac{15}{3}$$

$$x = 5$$

Q. Find the value of x in the equation $3 - (7x - 13) = 5(3 - x) - x$.

A. $x = 1$. Before you can begin rearranging terms, remove the parentheses on both sides of the equation. On the left side, the parentheses are preceded by a minus sign, so change the sign of every term and remove the parentheses:

$$3 - 7x + 13 = 5(3 - x) - x$$

On the right side, no sign comes between the 5 and the parentheses, so multiply every term inside the parentheses by 5 and remove the parentheses:

$$3 - 7x + 13 = 15 - 5x - x$$

Now you can solve the equation in three steps. Put the x terms on one side and the constants on the other, remembering to switch the signs as needed:

$-7x = 15 - 5x - x - 3 - 13$

$-7x + 5x + x = 15 - 3 - 13$

Combine like terms on both sides of the equation:

$-x = -1$

Divide by -1 to isolate x:

$$\frac{-x}{-1} = \frac{-1}{-1}$$
$$x = 1$$

---

9. Rearrange the equation $10x + 5 = 3x + 19$ to solve for x.

Solve It

10. Find the value of x by rearranging the equation $4 + (2x + 6) = 7(x - 5)$.

Solve It

---

11. Solve $-[2(x + 7) + 1] = x - 12$ for x.

Solve It

12. Find the value of x: $-x^3 + 2(x^2 + 2x + 1) = 4x^2 - (x^3 + 2x^2 - 18)$.

Solve It

Barring Fractions: Cross-Multiplying to Simplify Equations

Fraction bars, like parentheses, are grouping symbols: The numerator is one group, and the denominator is another. But like parentheses, fraction bars can block you from rearranging an equation and combining like terms. Luckily, cross-multiplication is a great trick for removing fraction bars from an algebraic equation.

You can use cross-multiplication to compare fractions, as I show you in Chapter 6. To show that fractions are equal, you can cross-multiply them — that is, multiply the numerator of one fraction by the denominator of the other. For example, here are two equal fractions. As you can see, when you cross-multiply them, the result is another balanced equation:

$$\frac{2}{5} = \frac{4}{10}$$
$$2(10) = 4(5)$$
$$20 = 20$$

You can use this trick to simplify algebraic equations that contain fractions.

Q. Use cross-multiplication to solve the equation $\frac{2x}{3} = x - 3$.

A. $x = 9$. Cross-multiply to get rid of the fraction in this equation. To do this, turn the right side of the equation into a fraction by inserting a denominator of 1:

$$\frac{2x}{3} = \frac{x-3}{1}$$

Now cross-multiply:

$$2x(1) = 3(x - 3)$$

Remove the parentheses from both sides (as I show you in Chapter 14):

$$2x = 3x - 9$$

At this point, you can rearrange the equation and solve for x, as I show you earlier in this chapter:

$$2x - 3x = -9$$
$$-x = -9$$
$$\frac{-x}{-1} = \frac{-9}{-1}$$
$$x = 9$$

Q. Use cross-multiplication to solve the equation $\frac{2x+1}{x+1} = \frac{6x}{3x+1}$.

A. $x = 1$. In some cases, after you cross-multiply, you may need to FOIL one or both sides of the resulting equation. First, cross-multiply to get rid of the fraction bar in this equation:

$$(2x + 1)(3x + 1) = 6x(x + 1)$$

Now remove the parentheses on the left side of the equation by FOILing (as I show you in Chapter 14):

$$6x^2 + 2x + 3x + 1 = 6x(x + 1)$$

To remove the parentheses from the right side, multiply $6x$ by every term inside the parentheses, and then drop the parentheses:

$$6x^2 + 2x + 3x + 1 = 6x^2 + 6x$$

At this point, you can rearrange the equation and solve for x:

$$1 = 6x^2 + 6x - 6x^2 - 2x - 3x$$

Notice that the two x^2 terms cancel each other out:

$$1 = 6x - 2x - 3x$$

$$1 = x$$

13. Rearrange the equation $\frac{x+5}{2} = \frac{-x}{8}$ to solve for x.

Solve It

14. Find the value of x by rearranging the equation $\frac{3x+5}{7} = x - 1$.

Solve It

15. Solve the equation $\frac{x}{2x-5} = \frac{2x+3}{4x-7}$.

Solve It

16. Find the value of x in this equation: $\frac{2x+3}{4-8x} = \frac{6-x}{4x+8}$.

Solve It

Answers to Problems in Finding the Right Balance: Solving Algebraic Equations

The following are the answers to the practice questions presented in this chapter.

1 Solve for x in each case just by looking at the equation.

a. $x + 5 = 13$; $x = 8$, because $8 + 5 = 13$.

b. $18 - x = 12$; $x = 6$, because $18 - 6 = 12$

c. $4x = 44$; $x = 11$, because $4(11) = 44$

d. $\frac{30}{x} = 3$; $x = 10$, because $\frac{30}{10} = 3$

2 Use the correct inverse operation to rewrite and solve each problem.

a. $x + 41 = 97$; $x = 56$. Change the addition to subtraction: $x + 41 = 97$ is the same as $97 - 41$, so $x = 56$.

b. $100 - x = 58$; $x = 42$. Change the subtraction to addition: $100 - x = 58$ means the same thing as $100 - 58 = x$, so $x = 42$.

c. $3x = 273$; $x = 21$. Change the multiplication to division: $13x = 273$ is equivalent to $\frac{273}{13} = x$, so $x = 21$.

d. $\frac{238}{x} = 17$; $x = 14$. Switch around the division: $\frac{238}{x} = 17$ means $\frac{238}{17} = x$, so $x = 14$.

3 Find the value of x in each equation by guessing and checking.

a. $19x + 22 = 136$; $x = 6$. First, try $x = 10$:

$19(10) + 22 = 190 + 22 = 212$

212 is greater than 136, so this guess is too high. Try $x = 5$:

$19(5) + 22 = 95 + 22 = 117$

117 is only a little less than 136, so this guess is a little too low. Try $x = 6$:

$19(6) + 22 = 114 + 22 = 136$

136 is correct, so $x = 6$.

b. $12x - 17 = 151$; $x = 14$. First, try $x = 10$:

$12(10) - 17 = 120 - 17 = 103$

103 is less than 151, so this guess is too low. Try $x = 20$:

$12(20) - 17 = 240 - 17 = 223$

223 is greater than 151, so this guess is too high. Therefore, x is between 10 and 20. Try $x = 15$:

$12(15) - 17 = 180 - 17 = 163$

163 is a little greater than 151, so this guess is a little too high. Try $x = 14$:

$12(14) - 17 = 168 - 17 = 151$

151 is correct, so $x = 14$.

c. $19x - 8 = 600$; $x = 32$. First, try $x = 10$:

$19(10) - 8 = 190 - 8 = 182$

182 is much less than 600, so this guess is much too low. Try $x = 30$:

$19(30) - 8 = 570 - 8 = 562$

562 is still less than 600, so this guess is still too low. Try $x = 35$:

$19(35) - 8 = 665 - 8 = 657$

657 is greater than 600, so this guess is too high. Therefore, x is between 30 and 35. Try $x = 32$:

$19(32) - 8 = 608 - 8 = 600$

600 is correct, so $x = 32$.

d. $x^2 + 3 = 292$; $\boldsymbol{x = 17}$. First, try $x = 10$:

$10^2 + 3 = 100 + 3 = 103$

103 is less than 292, so this guess is too low. Try $x = 20$:

$20^2 + 3 = 400 + 3 = 403$

403 is greater than 292, so this guess is too high. Therefore, x is between 10 and 20. Try $x = 15$:

$15^2 + 3 = 225 + 3 = 228$

228 is less than 292, so this guess is too low. Therefore, x is between 15 and 20. Try $x = 17$:

$17^2 + 3 = 289 + 3 = 292$

292 is correct, so $x = 17$.

4 Simplify the equation and then solve for x using any method you like:

a. $x^5 - 16 + x + 20 - x^5 = 24$; $\boldsymbol{x = 20}$. Rearrange the expression on the left side of the equation so that all like terms are next to each other:

$\underline{x^5 - x^5} + x + \underline{20 - 16} = 24$

Combine like terms:

$x + 4 = 24$

Notice that the two x^5 terms cancel each other out. Because $20 + 4 = 24$, you know that $x = 20$.

b. $5xy + x - 2xy + 27 - 3xy = 73$; $\boldsymbol{x = 46}$. Rearrange the expression on the left side of the equation:

$\underline{5xy - 2xy - 3xy} + x + 27 = 73$

Combine like terms:

$x + 27 = 73$

Notice that the three xy terms cancel each other out. Because $x + 27 = 73$ means $73 - 27 = x$, you know that $x = 46$.

c. $6x - 3 + x^2 - x + 8 - 5x = 30$; $\boldsymbol{x = 5}$. Rearrange the expression on the left side of the equation so that all like terms are adjacent:

$\underline{6x - x - 5x} + x^2 + \underline{8 - 3} = 30$

Combine like terms:

$x^2 + 5 = 30$

Notice that the three x terms cancel each other out. Try $x = 10$:

$10^2 + 5 = 100 + 5 = 105$

105 is greater than 30, so this guess is too high. Therefore, x is between 0 and 10. Try $x = 5$:

$$5^2 + 5 = 25 + 5 = 30$$

This result is correct, so $x = 5$.

d. $-3 + x^3 + 4 + x - x^3 - 1 = 2xy + 7 - x - 2xy + x$; $x = 7$. Rearrange the expression on the left side of the equation:

$$\underline{-3 + 4 - 1} + \underline{x^3 - x^3} + x = 2xy + 7 - x - 2xy + x$$

Combine like terms:

$$x = 2xy + 7 - x - 2xy + x$$

Notice that the three constant terms cancel each other out, and so do the two x^3 terms. Now rearrange the expression on the right side of the equation:

$$x = \underline{2xy - 2xy} + 7 \underline{- x + x}$$

Combine like terms:

$$x = 7$$

Notice that the two xy terms cancel each other out, and so do the two x terms. Therefore, $x = 7$.

5 $x = 3$. To get all constants on the right side of the equation, add 2 to both sides:

$$\begin{array}{r} 9x - 2 = 6x + 7 \\ \underline{+2 \qquad +2} \\ 9x \quad = 6x + 9 \end{array}$$

To get all x terms on the left side, subtract $6x$ from both sides:

$$\begin{array}{r} 9x \quad = \quad 6x + 9 \\ \underline{-6x \qquad -6x} \\ 3x \quad = \qquad 9 \end{array}$$

Divide by 3 to isolate x:

$$\frac{3x}{3} = \frac{9}{3}$$
$$x = 3$$

6 $x = 11$. Move all constants on the right side of the equation by adding 10 to both sides:

$$\begin{array}{r} 10x - 10 = 8x + 12 \\ \underline{+10 \qquad +10} \\ 10x \quad = 8x + 22 \end{array}$$

To get all x terms on the left side, subtract $8x$ from both sides:

$$10x = 8x + 22$$
$$\underline{-8x \quad -8x}$$
$$2x = 22$$

Divide by 2 to isolate x:

$$\frac{2x}{2} = \frac{22}{2}$$
$$x = 11$$

7 $x = 13$. Add 17 to both sides to get all constants on the right side of the equation:

$$4x - 17 = x + 22$$
$$\underline{+17 \qquad +17}$$
$$4x = x + 39$$

Subtract x from both sides to get all x terms on the left side:

$$4x = x + 39$$
$$\underline{-x \quad -x}$$
$$3x = 39$$

Divide by 3 to isolate x:

$$\frac{3x}{3} = \frac{39}{3}$$
$$x = 13$$

8 $x = 11$. To get all constants on the right side of the equation, add 40 to both sides:

$$15x - 40 = 11x + 4$$
$$\underline{+40 \qquad +40}$$
$$15x = 11x + 44$$

To get all x terms on the left side, subtract $11x$ from both sides:

$$15x = 11x + 44$$
$$\underline{-11x \quad -11x}$$
$$4x = 44$$

Divide by 4 to isolate x:

$$\frac{4x}{4} = \frac{44}{4}$$
$$x = 11$$

9 $x = 2$. Rearrange the terms of the equation so that the x terms are on one side and the constants are on the other. I do this in two steps:

$$10x + 5 = 3x + 19$$
$$10x = 3x + 19 - 5$$
$$10x - 3x = 19 - 5$$

Combine like terms on both sides:

$$7x = 14$$

Divide by 7 to isolate x:

$$\frac{7x}{7} = \frac{14}{7}$$
$$x = 2$$

10 $x = 9$. Before you can begin rearranging terms, remove the parentheses on both sides of the equation. On the left side, the parentheses are preceded by a plus sign, so just drop them:

$$4 + (2x + 6) = 7(x - 5)$$
$$4 + 2x + 6 = 7(x - 5)$$

On the right side, no sign comes between the number 7 and the parentheses, so multiply 7 by every term inside the parentheses and then drop the parentheses:

$$4 + 2x + 6 = 7x - 35$$

Now you can solve for x by rearranging the terms of the equation. Group the x terms on one side and the constants on the other. I do this in two steps:

$$4 + 6 = 7x - 35 - 2x$$
$$4 + 6 + 35 = 7x - 2x$$

Combine like terms on both sides:

$$45 = 5x$$

Divide by 5 to isolate x:

$$\frac{45}{5} = \frac{5x}{5}$$
$$9 = x$$

11 $x = -1$. Before you can begin rearranging terms, remove the parentheses on the left side of the equation. Start with the inner parentheses, multiplying 2 by every term inside that set:

$$-[2(x + 7) + 1] = x - 12$$
$$-[2x + 14 + 1] = x - 12$$

Next, remove the remaining parentheses, switching the sign of every term within that set:

$$-2x - 14 - 1 = x - 12$$

Now you can solve for x by rearranging the terms of the equation:

$$-2x - 14 - 1 + 12 = x$$
$$-14 - 1 + 12 = x + 2x$$

Combine like terms on both sides:

$$-3 = 3x$$

Divide by 3 to isolate x:

$$\frac{-3}{3} = \frac{3x}{3}$$
$$-1 = x$$

12 $x = 4$. Before you can begin rearranging terms, multiply the terms in the left-hand parentheses by 2 and remove the parentheses on both sides of the equation:

$$-x^3 + 2(x^2 + 2x + 1) = 4x^2 - (x^3 + 2x^2 - 18)$$
$$-x^3 + 2x^2 + 4x + 2 = 4x^2 - (x^3 + 2x^2 - 18)$$
$$-x^3 + 2x^2 + 4x + 2 = 4x^2 - x^3 - 2x^2 + 18$$

Rearrange the terms of the equation:

$$-x^3 + 2x^2 + 4x + 2 - 4x^2 + x^3 + 2x^2 = 18$$
$$-x^3 + 2x^2 + 4x - 4x^2 + x^3 + 2x^2 = 18 - 2$$

Combine like terms on both sides (notice that the x^3 and x^2 terms all cancel out):

$$4x = 16$$

Divide by 4 to isolate x:

$$\frac{4x}{4} = \frac{16}{4}$$
$$x = 4$$

13 $x = -4$. Remove the fraction from the equation by cross-multiplying:

$$\frac{x+5}{2} = \frac{-x}{8}$$
$$8(x+5) = -2x$$

Multiply to remove the parentheses from the left side of the equation:

$$8x + 40 = -2x$$

At this point, you can solve for x:

$$40 = -2x - 8x$$
$$40 = -10x$$
$$\frac{40}{-10} = \frac{-10x}{-10}$$
$$-4 = x$$

14 $x = 3$. Change the right side of the equation to a fraction by attaching a denominator of 1. Remove the fraction bar from the equation by cross-multiplying:

$$\frac{3x+5}{7} = \frac{x-1}{1}$$
$$3x + 5 = 7(x-1)$$

Multiply 7 by each term inside the parentheses to remove the parentheses from the right side of the equation:

$$3x + 5 = 7x - 7$$

Now solve for x:

$$5 = 7x - 7 - 3x$$
$$5 + 7 = 7x - 3x$$
$$12 = 4x$$
$$3 = x$$

15 $x = 5$. Remove the fractions from the equation by cross-multiplying:

$$\frac{x}{2x-5} = \frac{2x+3}{4x-7}$$
$$x(4x-7) = (2x+3)(2x-5)$$

Remove the parentheses from the left side of the equation by multiplying through by x; remove parentheses from the right side of the equation by FOILing:

$$4x^2 - 7x = 4x^2 - 10x + 6x - 15$$

Rearrange the equation:

$$4x^2 - 7x - 4x^2 + 10x - 6x = -15$$

Notice that the two x^2 terms cancel each other out:

$$-7x + 10x - 6x = -15$$
$$-3x = -15$$
$$\frac{-3x}{-3} = \frac{-15}{-3}$$
$$x = 5$$

16 $x = 0$. Remove the fractions from the equation by cross-multiplying:

$$\frac{2x+3}{4-8x} = \frac{6-x}{4x+8}$$

$$(2x+3)(4x+8) = (6-x)(4-8x)$$

FOIL both sides of the equation to remove the parentheses:

$$8x^2 + 16x + 12x + 24 = 24 - 48x - 4x + 8x^2$$

At this point, rearrange terms so you can solve for x:

$$8x^2 + 16x + 12x + 24 + 48x + 4x - 8x^2 = 24$$

$$8x^2 + 16x + 12x + 48x + 4x - 8x^2 = 24 - 24$$

Notice that the x^2 terms and the constant terms drop out of the equation:

$$16x + 12x + 48x + 4x = 0$$
$$80x = 0$$
$$\frac{80x}{80} = \frac{0}{80}$$
$$x = 0$$

Part V

The Part of Tens

For a list of ten great mathematicians, head to www.dummies.com/extras/ basicmathprealgebrawb.

In this part. . .

✔ Identify some early number systems, including Egyptian, Roman, and Mayan numerals.

✔ Understand prime numbers, including Mersenne and Fermat primes.

Chapter 16

Ten Alternative Numeral and Number Systems

In This Chapter

▶ Looking at numeral systems of the Egyptians, Babylonians, Romans, and Mayans

▶ Comparing the decimal number system with the binary and hexadecimal systems

▶ Taking a leap into the world of prime-based numbers

The distinction between numbers and numerals is subtle but important. A number is an idea that expresses how much or how many. A numeral is a written symbol that expresses a number.

In this chapter, I show you ten ways to represent numbers that differ from the Hindu-Arabic (decimal) system. Some of these systems use entirely different symbols from those you're used to; others use the symbols that you know in different ways. A few of these systems have useful applications, and the others are just curiosities. (If you like, you can always use them for sending secret messages!) In any case, you may find it fun and interesting to see how many different ways people have found to represent the numbers that you're accustomed to.

Tally Marks

Numbers are abstractions that stand for real things. The first known numbers came into being with the rise of trading and commerce — people needed to keep track of commodities such as animals, harvested crops, or tools. At first, traders used clay or stone tokens to help simplify the job of counting. The first numbers were probably an attempt to simplify this recording system. Over time, tally marks scratched either in bone or on clay took the place of tokens.

When you think about it, the use of tally marks over tokens indicates an increase in sophistication. Previously, one real object (a token) had represented another real object (for example, a sheep or ear of corn). After that, an *abstraction* (a scratch) represented a real object.

Bundled Tally Marks

As early humans grew more comfortable letting tally marks stand for real-world objects, the next development in numbers was probably tally marks scratched in *bundles* of 5 (fingers on one hand), 10 (fingers on both hands), or 20 (fingers and toes). Bundling provided a simple way to count larger numbers more easily.

Of course, this system is much easier to read than non-bundled scratches — you can easily multiply or count by fives to get the total. Even today, people keep track of points in games using bundles such as these.

Egyptian Numerals

Ancient Egyptian numerals are among the oldest number systems still in use today. Egyptian numerals use seven symbols, explained in Table 16-1.

Table 16-1	Egyptian Numerals
Number	**Symbol**
1	Stroke
10	Yoke
100	Coil of rope
1,000	Lotus
10,000	Finger
100,000	Frog
1,000,000	Man with raised hands

Numbers are formed by accumulating enough of the symbols that you need. For example,

7 = 7 strokes

24 = 2 yokes, 4 strokes

1,536 = 1 lotus, 5 coils of rope, 3 yokes, 6 strokes

In Egyptian numbers, the symbol for 1,000,000 also stands for infinity (∞).

Babylonian Numerals

Babylonian numerals, which came into being about 4,000 years ago, use two symbols:

1 = Y   10 = <

For numbers less than 60, numbers are formed by accumulating enough of the symbols you need. For example,

6 = YYYYYY

34 = <<<YYYY

59 = <<<<<YYYYYYYYY

For numbers 60 and beyond, Babylonian numerals use place value based on the number 60. For example,

61 = Y Y	(one 60 and one 1)
124 = YY YYYY	(two 60s and four 1s)
611 = < <Y	(ten 60s and eleven 1s)

Unlike the decimal system that you're used to, Babylonian numbers had no symbol for zero to serve as a placeholder, which causes some ambiguity. For example, the symbol for 60 is the same as the symbol for 1.

Ancient Greek Numerals

Ancient Greek numerals were based on the Greek letters. The numbers from 1 to 999 were formed using the symbols in Table 16-2.

Table 16-2	Numerals Based on the Greek Alphabet	
Ones	*Tens*	*Hundreds*
1 = α (alpha)	10 = ι (iota)	100 = ρ (rho)
2 = β (beta)	20 = κ (kappa)	200 = σ (sigma)
3 = γ (gamma)	30 = λ (lambda)	300 = τ (tau)
4 = δ (delta)	40 = μ (mu)	400 = υ (upsilon)
5 = ε (epsilon)	50 = ν (nu)	500 = φ (phi)
6 = ς (digamma)	60 = ξ (xi)	600 = χ (chi)
7 = ζ (zeta)	70 = o (omicron)	700 = ψ (psi)
8 = η (eta)	80 = π (pi)	800 = ω (omega)
9 = θ (theta)	90 = ϟ (koppa)	900 = ϡ (sampi)

Roman Numerals

Although Roman numerals are over 2,000 years old, people still use them today, either decoratively (for example, on clocks, cornerstones, and Super Bowl memorabilia) or when numerals distinct from decimal numbers are needed (for example, in outlines). Roman numerals use seven symbols, all of which are capital letters in the Latin alphabet (which pretty much happens to be the English alphabet as well):

I = 1   V = 5   X = 10   L = 50
C = 100   D = 500   M = 1,000

Most numbers are formed by accumulating enough of the symbol that you need. Generally, you list the symbols in order, from highest to lowest. Here are a few examples:

3 = III   8 = VIII   20 = XX   70 = LXX
300 = CCC   600 = DC   2,000 = MM

Numbers that would contain 4s or 9s in the decimal system are formed by transposing two numbers to indicate subtraction. When you see a smaller symbol come before a larger one, you have to subtract the smaller value from the number that comes after it:

4 = IV   9 = IX   40 = XL
90 = XC   400 = CD   900 = CM

These two methods of forming numbers are sufficient to represent all decimal numbers up to 3,999:

37 = XXXVII   664 = DCLXIV
1,776 = MDCCLXXVI   1,999 = MCMXCIX

Higher numbers are less frequent, but you form them by putting a bar over a symbol, beginning with a bar over V for 5,000 and ending with a bar over M for 1,000,000. The bar means that you need to multiply by 1,000.

Mayan Numerals

Mayan numerals developed in South America during roughly the same period that Roman numerals developed in Europe. Mayan numerals use two symbols: dots and horizontal bars. A bar is equal to 5, and a dot is equal to 1. Numbers from 1 to 19 are formed by accumulating dots and bars. For example,

  3 = 3 dots

  7 = 2 dots over 1 bar

  19 = 4 dots over 3 bars

Numbers from 20 to 399 are formed using these same combinations, but raised up to indicate place value. For example,

  21 = raised 1 dot, 1 dot (one 20 + one 1)

  86 = raised 4 dots, 1 dot over 1 bar (four 20s + one 5 + one 1)

  399 = raised 4 dots over 3 bars, 4 dots over 3 bars (nineteen 20s + three 5s + four 1s)

As you can see, Mayan place value is based on the number 20 rather than the number 10 that we use. Numbers from 400 to 7,999 are formed similarly, with an additional place — the 400s place.

Because Mayan numerals use place value, there's no limit to the magnitude of numbers that you can express. This fact makes Mayan numerals more mathematically advanced than either Egyptian or Roman numerals. For example, you could potentially use Mayan numerals to represent astronomically large numbers — such as the number of stars in the known universe or the number of atoms in your body — without changing the basic rules of the system. This sort of representation would be impossible with Egyptian or Roman numerals.

Base-2 (Binary) Numbers

Binary numbers use only two symbols: 0 and 1. This simplicity makes binary numbers useful as the number system that computers use for data storage and computation.

Like the decimal system you're most familiar with, binary numbers use place value (see Chapter 1 for more on place value). Unlike the decimal system, binary place value is based not on powers of ten (1, 10, 100, 1,000, and so forth) but on powers of two (2^0, 2^1, 2^2, 2^3, 2^4, 2^5, 2^6, 2^7, 2^8, 2^9, and so on), as seen in Table 16-3 (see Chapter 2 for more on powers).

Table 16-3				Binary Place Values					
512s	256s	128s	64s	32s	16s	8s	4s	2s	1s

Notice that each number in the table is exactly twice the value of the number to its immediate right. Note also that the base-2 number system is based on the *base* of a bunch of exponents (see Chapter 2, which covers powers). You can use this chart to find out the decimal value of a binary number. For example, suppose you want to represent the binary number 1101101 as a decimal number. First, place the number in the binary chart, as in Table 16-4.

Table 16-4				Breaking Down a Binary Number					
512s	256s	128s	64s	32s	16s	8s	4s	2s	1s
			1	1	0	1	1	0	1

The table tells you that this number consists of one 64, one 32, no 16s, one 8, one 4, no 2s, and one 1. Add these numbers up, and you find that the binary number 1101101 equals the decimal number 109:

$$64 + 32 + 8 + 4 + 1 = 109$$

To translate a decimal number into its binary equivalent, use whole-number division to get a quotient and a remainder (as I explain in Chapter 1). Start by dividing the number you're translating into the next-highest power of 2. Keep dividing powers of 2 into the remainder of the result. For example, here's how to find out how to represent the decimal number 83 as a binary number:

$$83 \div 64 = \mathbf{1} \text{ r } 19$$

$$19 \div 32 = \mathbf{0} \text{ r } 19$$

$$19 \div 16 = \mathbf{1} \text{ r } 3$$

$$3 \div 8 = \mathbf{0} \text{ r } 3$$

$$3 \div 4 = \mathbf{0} \text{ r } 3$$

$$3 \div 2 = \mathbf{1} \text{ r } 1$$

$$1 \div 1 = \mathbf{1} \text{ r } 0$$

So the decimal number 83 equals the binary number 1010011 because 64 + 16 + 2 + 1 = 83.

Base-16 (Hexadecimal) Numbers

The computer's first language is binary numbers. But in practice, humans find binary numbers of any significant length virtually undecipherable. Hexadecimal numbers, however, are readable to humans and still easily translated into binary numbers, so computer programmers use hexadecimal numbers as a sort of common language when interfacing with computers at the deepest level, the level of hardware and software design.

The hexadecimal number system uses all ten digits 0 through 9 from the decimal system. Additionally, it uses six more symbols:

A = 10	B = 11	C = 12
D = 13	E = 14	F = 15

Hexadecimal is a place-value system based on powers of 16, as shown in Table 16-5.

Table 16-5		Hexadecimal Place Values			
1,048,576s	65,536s	4,096s	256s	16s	1s

As you can see, each number in the table is exactly 16 times the number to its immediate right.

Here are a few examples of hexadecimal numbers and their equivalent representations in decimal notation:

$$3B = (3 \times 16) + 11 = 59$$

$$289 = (2 \times 256) + (8 \times 16) + 9 = 649$$

$$ABBA = (10 \times 4,096) + (11 \times 256) + (11 \times 16) + 10 = 43,962$$

$$B00B00 = (11 \times 1,048,576) + (11 \times 256) = 11,537,152$$

Prime-Based Numbers

One wacky way to represent numbers unlike any of the others in this chapter is prime-based numbers. Prime-based numbers are similar to decimal, binary, and hexadecimal numbers (which I describe earlier) in that they use place value to determine the value of digits. But unlike these other number systems, prime-based numbers are based not on addition but on multiplication. Table 16-6 shows a place value chart for prime-based numbers.

Table 16-6						Prime-Based Place Values				
31s	29s	23s	19s	17s	13s	11s	7s	5s	3s	2s

You can use the table to find the decimal value of a prime-based number. For example, suppose you want to represent the prime-based number 1,204 as a decimal number. First, place the number in the table, as shown in Table 16-7.

Table 16-7					Breaking Down a Prime-Based Number					
31s	29s	23s	19s	17s	13s	11s	7s	5s	3s	2s
							1	2	0	4

As you may have guessed, the table tells you that this number consists of one 7, two 5s, no 3s, and four 2s. But instead of adding these numbers together, you *multiply* them:

$$7 \times 5 \times 5 \times 2 \times 2 \times 2 \times 2 = 2,800$$

To translate a decimal number into its prime-based equivalent, factor the number and place its factors into the chart. For example, suppose you want to represent the decimal number 60 as a prime-based number. First, decompose 60 into its prime factors (as I show you in Chapter 5):

$$60 = 2 \times 2 \times 3 \times 5$$

Now count the number of twos, threes, and fives and place these in Table 16-6. The result should look like Table 16-8.

Table 16-8					Finding the Prime-Based Equivalent of 60					
31s	29s	23s	19s	17s	13s	11s	7s	5s	3s	2s
								1	1	2

So, 60 in prime-based notation is 112.

Interestingly, multiplication with prime-based numbers looks like addition with decimal numbers. For example, in decimal numbers, $9 \times 10 = 90$. The prime-based equivalents of the factors and product — 9, 10, and 90 — are 20, 101, and 121. So here's how to do the same multiplication in prime-based numbers:

$$20 \times 101 = 121$$

As you can see, this multiplication looks more like addition. Even weirder is that 1 in decimal notation is represented as 0 in prime-based notation. This makes sense when you think about it, because multiplying by 1 is very similar to adding 0.

Chapter 17

Ten Curious Types of Numbers

*N*umbers seem to have personalities all their own. For example, even numbers are go-along numbers that break in half so you can carry them more conveniently. Odd numbers are more stubborn and don't break apart so easily. Powers of ten are big friendly numbers that are easy to add and multiply, whereas most other numbers are prickly and require special attention. In this chapter, I introduce you to some interesting types of numbers, with properties that other numbers don't share.

Square Numbers

When you multiply any number by itself, the result is a *square number*. For example,

$$1^2 = 1 \times 1 = 1$$
$$2^2 = 2 \times 2 = 4$$
$$3^2 = 3 \times 3 = 9$$
$$4^2 = 4 \times 4 = 16$$
$$5^2 = 5 \times 5 = 25$$

Therefore, the sequence of square numbers begins as follows:

1, 4, 9, 16, 25, . . .

To see why they're called square numbers, look at the arrangement of coins in squares in Figure 17-1.

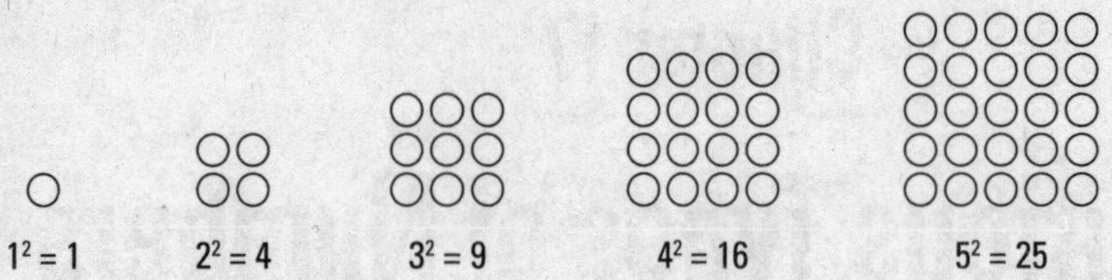

Figure 17-1:
The first
five square
numbers.

$1^2 = 1$ $2^2 = 4$ $3^2 = 9$ $4^2 = 16$ $5^2 = 25$

What's really cool about the list of square numbers is that you can get it by adding the odd numbers (3, 5, 7, 9, 11, 13, . . .), beginning with 3, to each preceding number in the list:

Square Number	Preceding Number + Odd Number	Sum
$2^2 = 4$	$1^2 + 3$	$1 + 3 = 4$
$3^2 = 9$	$2^2 + 5$	$4 + 5 = 9$
$4^2 = 16$	$3^2 + 7$	$9 + 7 = 16$
$5^2 = 25$	$4^2 + 9$	$16 + 9 = 25$
$6^2 = 36$	$5^2 + 11$	$25 + 11 = 36$
$7^2 = 49$	$6^2 + 13$	$36 + 13 = 49$

Triangular Numbers

When you add up any sequence of consecutive positive numbers starting with 1, the result is a *triangular number*. For example,

$$1 = 1$$
$$1 + 2 = 3$$
$$1 + 2 + 3 = 6$$
$$1 + 2 + 3 + 4 = 10$$
$$1 + 2 + 3 + 4 + 5 = 15$$

So, the sequence of triangular numbers begins as follows:

1, 3, 6, 10, 15, . . .

Triangular numbers' shapely name makes sense when you begin arranging coins in triangles. Check out Figure 17-2.

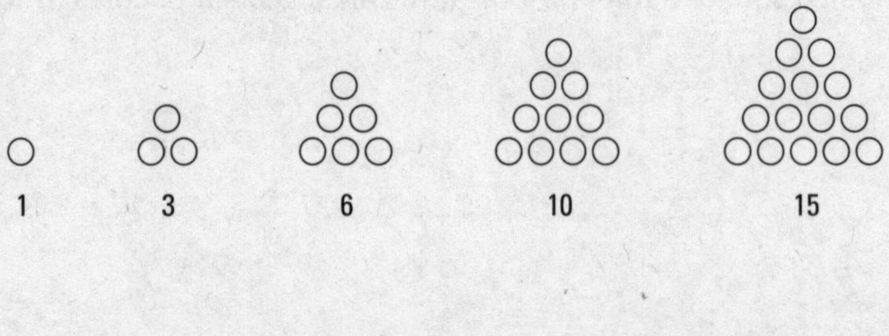

Figure 17-2:
The first five
triangular
numbers.

1   3   6   10   15

Cubic Numbers

If you're feeling that the square and triangular numbers are too flat, add a dimension and begin playing with the *cubic numbers*. You can generate a cubic number by multiplying any number by itself three times:

$$1^3 = 1 \times 1 \times 1 = 1$$
$$2^3 = 2 \times 2 \times 2 = 8$$
$$3^3 = 3 \times 3 \times 3 = 27$$
$$4^3 = 4 \times 4 \times 4 = 64$$
$$5^3 = 5 \times 5 \times 5 = 125$$

The sequence of cubic numbers begins as follows:

1, 8, 27, 64, 125, . . .

Cubic numbers live up to their name. Look at the cubes in Figure 17-3.

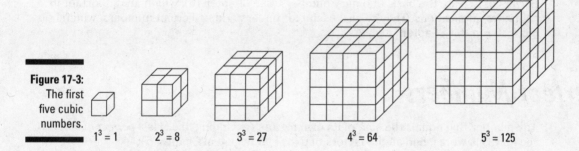

Figure 17-3: The first five cubic numbers.

$1^3 = 1$ $2^3 = 8$ $3^3 = 27$ $4^3 = 64$ $5^3 = 125$

Factorial Numbers

In math, the exclamation point (!) means *factorial*, so you read 1! as *one factorial*. You get a factorial number when you multiply any sequence of consecutive positive numbers, starting with the number itself and counting down to 1. For example,

$$1! = 1$$
$$2! = 2 \times 1 = 2$$
$$3! = 3 \times 2 \times 1 = 6$$
$$4! = 4 \times 3 \times 2 \times 1 = 24$$
$$5! = 5 \times 4 \times 3 \times 2 \times 1 = 120$$

Thus, the sequence of factorial numbers begins as follows:

1, 2, 6, 24, 120, ...

Factorial numbers are very useful in *probability*, which is the mathematics of how likely an event is to occur. With probability problems, you can figure out how likely you are to win the lottery or estimate your chances of guessing your friend's locker combination within the first few tries.

Powers of Two

Multiplying the number 2 by itself repeatedly gives you the *powers of two*. For example,

$2^1 = 2$

$2^2 = 2 \times 2 = 4$

$2^3 = 2 \times 2 \times 2 = 8$

$2^4 = 2 \times 2 \times 2 \times 2 = 16$

$2^5 = 2 \times 2 \times 2 \times 2 \times 2 = 32$

Powers of two are the basis of binary numbers (see Chapter 16), which are important in computer applications. They're also useful for understanding Fermat numbers, which I discuss later in this chapter.

Perfect Numbers

Any number that equals the sum of its own factors (excluding itself) is a *perfect number*. To see how this works, find all the factors of 6 (as I show you in Chapter 5):

6: 1, 2, 3, 6

Now add up all these factors except 6:

1 + 2 + 3 = 6

These factors add up to the number you started with, so 6 is a perfect number. The next perfect number is 28. First, find all the factors of 28:

28: 1, 2, 4, 7, 14, 28

Now add up all these factors except 28:

1 + 2 + 4 + 7 + 14 = 28

Again, these factors add up to the number you started with, so 28 is a perfect number. Perfect numbers are few and far between. The sequence of perfect numbers begins with the following five numbers:

6; 28; 496; 8,128; 33,550,336; . . .

You can use the same method I outline to check 496 and beyond by yourself.

Amicable Numbers

Amicable numbers are similar to perfect numbers, except they come in pairs. The sum of the factors of one number (excluding the number itself) is equal to the second number, and vice versa. For example, one amicable pair is 220 and 284. To see why, first find all the factors of each number:

220: 1, 2, 4, 5, 10, 11, 20, 22, 44, 55, 110, 220

284: 1, 2, 4, 71, 142, 284

For each number, add up all the factors except the number itself:

$1 + 2 + 4 + 5 + 10 + 11 + 20 + 22 + 44 + 55 + 110 = 284$

$1 + 2 + 4 + 71 + 142 = 220$

Notice that the factors of 220 add up to 284, and the factors of 284 add up to 220. That's what makes this pair of numbers amicable.

The next-lowest pair of amicable numbers is 1,184 and 1,210. You can either trust me on this one or do the calculation yourself.

Prime Numbers

Any number that has exactly two factors — 1 and itself — is called a *prime number*. For example, here are the first few prime numbers:

2, 3, 5, 7, 11, 13, 17, 19, . . .

There are infinitely many prime numbers — that is, they go on forever. See Chapter 5 for more on prime numbers.

Mersenne Primes

Any number that's 1 less than a power of two (which I discuss earlier in this chapter) is called a *Mersenne number* (named for French mathematician Marin Mersenne). Therefore, every Mersenne number is of the following form:

$2^n - 1$ (where n is a nonnegative integer)

When a Mersenne number is also a prime number (see the preceding section), it's called a *Mersenne prime*. For example,

$2^2 - 1 = 4 - 1 = 3$

$2^3 - 1 = 8 - 1 = 7$

$2^5 - 1 = 32 - 1 = 31$

$2^7 - 1 = 128 - 1 = 127$

$2^{13} - 1 = 8,192 - 1 = 8,191$

Mersenne primes are of interest to mathematicians because they possess properties that ordinary prime numbers don't have. One of these properties is that they tend to be easier to find than other prime numbers. For this reason, the search for the largest known prime number is usually a search for a Mersenne prime.

Fermat Primes

A *Fermat number* (named for mathematician Pierre de Fermat) is of the following form:

$2^{2^n} + 1$ (where n is a nonnegative integer)

The ^ symbol means that you're finding a power, so with this formula, you first find 2^n; then you use that answer as an exponent on 2. For example, here are the first five Fermat numbers:

$2^{2^0} + 1 = 2^1 + 1 = 3$

$2^{2^1} + 1 = 2^2 + 1 = 5$

$2^{2^2} + 1 = 2^4 + 1 = 16 + 1 = 17$

$2^{2^3} + 1 = 2^8 + 1 = 256 + 1 = 257$

$2^{2^4} + 1 = 2^{16} + 1 = 65,536 + 1 = 65,537$

As you can see, Fermat numbers grow very quickly. When a Fermat number is also a prime number (see earlier in this chapter), it's called a *Fermat prime*. As it happens, the first five Fermat numbers are also Fermat primes (testing this is fairly simple for the first four numbers above and *much* harder for the fifth).

Index